T0345272

Green Machine Learning Protocols for Future Communication Networks

Machine learning has shown tremendous benefits in solving complex network problems and providing situation and parameter prediction. However, heavy resources are required to process and analyze the data, which can be done either offline or using edge computing but also requires heavy transmission resources to provide a timely response. The need here is to provide lightweight machine learning protocols that can process and analyze the data at run time and provide a timely and efficient response. These algorithms have grown in terms of computation and memory requirements due to the availability of large data sets. These models/algorithms also require high levels of resources such as computing, memory, communication, and storage. The focus so far was on producing highly accurate models for these communication networks without considering the energy consumption of these machine learning algorithms.

For future scalable and sustainable network applications, efforts are required toward designing new machine learning protocols and modifying the existing ones, which consume less energy, i.e., green machine learning protocols. In other words, novel and lightweight green machine learning algorithms/protocols are required to reduce energy consumption which can also reduce the carbon footprint. To realize the green machine learning protocols, this book presents different aspects of green machine learning for future communication networks. This book highlights mainly the green machine learning protocols for cellular communication, federated learning-based models, and protocols for Beyond Fifth Generation networks, approaches for cloud-based communications, and Internet-of-Things. This book also highlights the design considerations and challenges for green machine learning protocols for different future applications.

Green Machine Learning Protocols for Future Communication Networks

Edited by

Saim Ghafoor
Mubashir Husain Rehmani

CRC Press is an imprint of the
Taylor & Francis Group, an **informa** business

Cover Image Credit: Shutterstock

First edition published 2024
by CRC Press
2385 Executive Center Drive, Suite 320, Boca Raton, FL 33431

and by CRC Press
4 Park Square, Milton Park, Abingdon, Oxon, OX14 4RN

CRC Press is an imprint of Taylor & Francis Group, LLC

ISBN: 978-1-032-13685-1 (hbk)
ISBN: 978-1-032-13687-5 (pbk)
ISBN: 978-1-003-23042-7 (ebk)

DOI: 10.1201/9781003230427

Typeset in Sabon
by SPi Technologies India Pvt Ltd (Straive)

Contents

Preface

Mobile communication technologies are already in their fast development phase and mobile data traffic has been forecasted to reach 226 exabytes per month by 2026 (Cisco). This requires high-quality communication network services and resources to fill the gap between the future requirements of wireless services and existing communication technologies. This gap can also be filled by designing intelligent algorithms and schemes to efficiently utilize the limited wireless resources. Machine learning (ML), in this regard, can play a vital role. For example, it can be used for analysis and prediction in wireless communication systems such as spectrum utilization, channel capacity, antenna configurations, and link selection. Machine learning has been used recently in many communication networks like wireless sensor networks, Internet-of-Things, cognitive radio networks, satellite communication, cloud/edge computing, software-defined networking, and machine-to-machine networks.

Machine learning has shown tremendous benefits in solving complex network problems and providing situation and parameter prediction. However, heavy resources are required to process and analyze the data which can be done either offline or using edge computing, which also requires heavy transmission rates to provide a timely response. The need here is to provide lightweight ML protocols that can process and analyze the data at run time and provide a timely and efficient response. The focus so far was on producing highly accurate models for these communication networks without considering the energy consumption of these machine learning algorithms. These algorithms have grown in terms of computation and memory requirements due to the availability of large data sets. These models/algorithms also require high levels of resources such as computing, memory, communication, and storage.

Little effort has been made so far on reducing the energy consumption of these models and algorithms. For future scalable and sustainable network applications, efforts are required toward designing new machine learning protocols and modifying the existing ones, which consume less energy, i.e., green machine learning protocols. In other words, novel and lightweight green machine learning algorithms/protocols are required to reduce energy consumption which can also reduce the carbon footprint.

There are many books on machine learning, deep learning, and federated learning and their modeling and accuracy. However, none of the books are discussing the green effect requirement of machine learning algorithms, their significance, and their suitability for future 6th Generation (6G) communication. This is the first book presenting the green machine learning communication aspects for applications like cellular communication, cloud and data center communication, and Internet-of-Things. This book also highlights the challenges for different communication networks for 6G communications using Green Machine Learning mechanisms.

The contents of this book show the relevancy of the book for researchers at different levels and postgraduate students who are looking to enhance their knowledge and understanding of using Green Machine Learning mechanisms for different communication networks.

Intended audience

This book would be an interest to multiple groups of researchers, postgraduate students, data science experts, and network operators, who are working on application and protocols development and deployment.

- **Academics:** Machine learning has already been a major area of research and study for many educational institutions all over the world. This book can be an essential part of academics in understanding this interesting field and also in enhancing and addressing the research challenges in designing energy-efficient machine learning protocols and algorithms.
- **Network operators:** The network operators will be looking to adopt B5G and 6G technology to offer new services to their customers. The important part of these energy-efficient machine learning aspects is to explore the architecture of machine learning-based applications, as they are mostly moving to edge networks. This book will provide guidelines, and techniques to deploy and develop machine learning-based applications onto the cloud.
- **Data Center Operators:** This book will be helpful for cloud and data center operatives in managing the resources of resource-heavy machine learning-based applications.
- **Telecommunication researchers:** Machine learning is one of the key interests for telecommunication researchers for the high capacity and flexible user services availability. This book will offer a great source of major green machine learning protocols design and working challenges of cellular communications.
- **Internet-of-things:** This is one major application which focuses more on energy-efficient protocols while maintaining latency. This book will help researchers working in the Internet-of-Things field in addressing the challenges of lightweight machine learning algorithms for the Internet-of-Things.

Book organization

This book covers various aspects of green machine learning protocols for different communication networks like cellular communication, cloud and data center networks, Internet-of-Things, and machine-to-machine networks. This book also highlights the challenges of lightweight protocols for machine learning for these networks. Mainly, this book contains seven chapters covering different aspects of green machine learning for these networks.

In Chapter 1, the significance of green machine learning protocols for cellular networks is discussed in contrast with computationally expensive deep-learning algorithms. The requirements of green machine learning algorithms for cellular networks are discussed and design challenges are highlighted for optimized solutions. In addition, several new performance metrics and their performance evaluation are also mentioned.

Chapter 2 takes the discussion on cellular communication forward and discusses the state-of-the-art use of green machine learning approaches in solving problems and expansion of cellular communication networks and ad hoc networks. At first, the chapter discusses the quality-of-service constraints, including effective capacity, energy consumption, spectral efficiency, low power regime energy efficiency, wideband system energy efficiency, fixed rate and power transmissions, and training and data transmission phases. Secondly, this chapter discusses the trade-offs of energy spectral efficiency for green machine learning in cellular networks. This chapter also covers the highlights and challenges for cellular wireless base stations and access network energy efficiency including network design, deployment, user association, and base station active sleep patterns. In the end, this chapter highlights and discusses many open issues and challenges for green cellular networks, including green machine learning approaches challenges for cellular communications, and presents a management model for green wireless networks.

Chapter 3 introduces federated learning which is a key technique for collaboratively training machine learning models while aggregating data on a centralized entity. This chapter initially discusses different considerations for green federated learning models and protocols. This chapter identifies and highlights the key enablers (including hardware and software), techniques, and their usage to achieve green federated learning. It provides an overview of efforts taken until now for federated learning solutions and identifies their shortcomings. It also discusses lightweight model design considerations. It also highlights many problems and challenges of current design in the development of green federated learning models and protocols.

In Chapter 4 the energy-efficient and privacy-protecting distributed learning techniques in communication networks are discussed. The chapter also highlights the problem of data collection and processing and discusses federated learning to effectively train ground-truth data on-device with only model update parameters sent back to the federated server. Typically, in this chapter, a chameleon federated learning framework based green 6G model is proposed for network slicing in Beyond 5G systems to solve complex load estimation problems without the need of collecting sensitive private data from end devices.

In Chapter 5, an overview of green cloud computing is presented. Mainly, this chapter provides a comprehensive literature review of existing methods and approaches to improve energy utilization in cloud computing. Machine learning approaches for cloud computing are discussed emphasizing energy usage, consumption, and efficiency with resource management. It also discusses deepMind and AI-based solutions for cloud data centers. This chapter also documents the future trends and challenges for green machine learning-based cloud data centers.

Chapter 6 highlights the limitation of Internet-of-Things devices such as limited storage and energy efficiency requirements while machine learning algorithms and models usage. This chapter also highlights the need for green machine learning architectures, models, and solutions. Mainly, this chapter provides a comprehensive survey of the application of machine learning in the Internet-of-Things domain. It discusses different solutions for the usage of green machine learning for Internet-of-Things and also presents challenges and future trends.

Chapter 7 discusses the random-access channel for massive machine-to-machine communication and highlights its problems with the extended access baring mechanism proposed in the 3rd Generation Partnership Project. This chapter presents the green machine learning approaches and their necessities for massive machine-to-machine communication for 5G cellular Internet-of-Things. The future trends and challenges are also presented for massive machine-to-machine green machine learning communication.

Acknowledgments

First, we are thankful to Allah Subhanahu Wa-ta'ala that by His grace and bounty we were able to complete the editing of this book.

We would like to thank all the chapter authors for their hard work and involvement from the beginning till the finishing of this book. We would also like to thank our reviewers for their voluntarily work. They helped a lot in improving the overall quality of this book.

We wish to express gratitude to Sheikh Hazrat Mufti Mohammad Naeem Memon of Hyderabad, Pakistan. We could not have edited this book without his prayers, spiritual guidance, and moral support.

Last but not the least, we would like to thank our parents and families for their continuous love, motivation, and prayers.

Saim Ghafoor
Atlantic Technological University,
Letterkenny, Ireland
February 2, 2023

Mubashir Husain Rehmani
Munster Technological University,
Cork, Ireland
February 2, 2023

Editors

Saim Ghafoor received his PhD in computer science from University College Cork, Ireland in 2018. He completed his MS in Computer Science and Engineering from Hanyang University, South Korea in 2010 and BE in Computer Systems Engineering from Mehran University of Engineering and Technology, Pakistan in 2005. Currently, he is working as lecturer at Atlantic Technological University (ATU), Donegal, Ireland. He worked at Telecommunications Software and Systems Group (TSSG), Waterford Institute of Technology (WIT), Waterford, Ireland as a post-doctoral researcher from June 2018 to August 2020. He is currently an associate editor of the Elsevier *Computer and Electrical Engineering Journal* and has served as a reviewer for many reputable journals and conferences. He co-organized a workshop on Terahertz wireless communication in IEEE INFOCOM 2019. He also received the best student paper award in International Conference on Disaster Management in 2017. He received the best reviewer of the year award from *Elsevier Computer and Electrical Engineering Journal* in 2014 and 2015, and many reviewer recognitions certificates. His research interests are Terahertz and millimeter communication networks, autonomous and intelligent communication networks, cognitive radio networks, and wireless sensor networks.

Mubashir Husain Rehmani received the BE degree in Computer Systems Engineering from Mehran University of Engineering and Technology, Jamshoro, Pakistan; in 2004, the MS degree from the University of Paris XI, Paris, France, in 2008; and the PhD degree from the University Pierre and Marie Curie, Paris, in 2011. He is currently working as a lecturer in the Department of Computer Science, Munster Technological

University (MTU), Ireland. Prior to this, he worked as a post-doctoral researcher at the Telecommunications Software and Systems Group (TSSG), Waterford Institute of Technology (WIT), Waterford, Ireland. He also served for five years as an assistant professor at COMSATS Institute of Information Technology, Wah Cantt., Pakistan. He is serving as an Editorial Board Member of *NATURE Scientific Reports*. He is currently an area editor of the *IEEE Communications Surveys and Tutorials*. He served for three years (from 2015 to 2017) as an associate editor of the *IEEE Communications Surveys and Tutorials*. He served as a column editor for Book Reviews in *IEEE Communications Magazine*. He is appointed as an associate editor for *IEEE Transactions on Green Communication and Networking*. Currently, he serves as an associate editor of *IEEE Communications Magazine*, Elsevier *Journal of Network and Computer Applications (JNCA)*, and the *Journal of Communications and Networks (JCN)*. He is also serving as a guest editor of Elsevier *Ad Hoc Networks* journal, Elsevier *Future Generation Computer Systems* journal, the *IEEE Transactions on Industrial Informatics*, and Elsevier *Pervasive and Mobile Computing* journal. He has authored/edited total eight books. Two books with Springer, two books published by IGI Global, USA, three books published by CRC Press – Taylor and Francis Group, UK, and one book with Wiley, UK. He received the "Best Researcher of the Year 2015 of COMSATS Wah" award in 2015. He received the certificate of appreciation, "*Exemplary Editor of the IEEE Communications Surveys and Tutorials for the year 2015*" from the IEEE Communications Society. He received the Best Paper Award from IEEE ComSoc Technical Committee on Communications Systems Integration and Modeling (CSIM), in IEEE ICC 2017. He consecutively received research productivity award in 2016–17 and also ranked # 1 in all Engineering disciplines from Pakistan Council for Science and Technology (PCST), Government of Pakistan. He received the Best Paper Award in 2017 from Higher Education Commission (HEC), Government of Pakistan. He is recipient of the Best Paper Award in 2018 from *Elsevier Journal of Network and Computer Applications*. He is a three-time recipient of the *Highly Cited Researcher*™ award—in 2020, 2021, and 2022—by Clarivate, USA. His performance in this context features in the *TOP 1%* by citations in the field of computer science and Cross Field in the Web of Science™ citation index. He is the only researcher from Ireland in the field of "Computer Science" who received this international prestigious award. In October 2022, he received Science Foundation Ireland's CONNECT Centre's Education and Public Engagement (EPE) Award 2022 for his research outreach work and being a spokesperson for achieving a work–life balance for a career in research.

Contributors

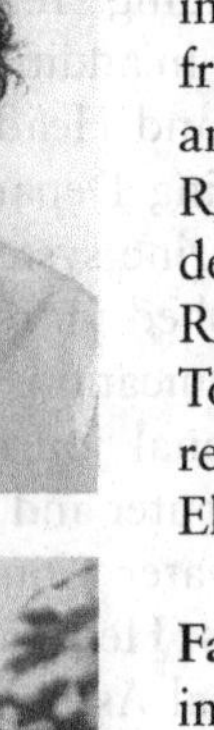

Amine Abouaomar is currently an assistant professor at Al Akhawayn University in Ifrane, Morocco. Dr. Abouaomar received his PhD. in electrical engineering and computer science from Université of Sherbrooke (QC) Canada, and ENSIAS, Mohammed V University of Rabat, Morocco in 2021. He received the B.Sc. degree in computer science and the M.Sc. of Research degree in computer science from Ibn Tofail University, Morocco, in 2014 and 2016, respectively. Dr. Abouoamar received the Mitacs Elevate Fellowship in 2022.

Fakhrul Alam received the BSc degree (Hons.) in electrical and electronic engineering from BUET, Bangladesh, and the MS and PhD degrees in electrical engineering from Virginia Tech, USA. He is currently an associate professor in the Department of Mechanical and Electrical Engineering (MEE), School of Food and Advanced Technology, Massey University, New Zealand. He is the Department Lead-MEE, and Subject Lead-Engineering of Massey University. He also holds the position of adjunct professor with the School of Engineering and Technology of Sunway University, Malaysia, for 2021–22. His research involves the development of Intelligent Systems, Smart Sensors, and Precision Instrumentation by leveraging his expertise in Wireless and Visible Light Communication, IoT and Signal Processing. His research work has received $1M+ external funding (as PI) from the likes of the New Zealand Ministry of Business, Innovation and Employment (MBIE), Auckland Transport and New Zealand Forest Research Institute Limited. He is a senior member of the Institute of Electrical & Electronics Engineers

(IEEE), and a member of the Association for Computing Machinery (ACM). He is an associate editor of *IEEE Access*, topic editor for *Sensors (MDPI)*, sits on the IEEE Conference Technical Program Integrity Committee (TPIC), and serves as a reviewer for the Royal Society Te Apārangi. He is a regular invited speaker at various national and international fora and has led (e.g., as General/TPC/Program/OC chair) the organization of many international conferences.

Elmustafa Sayed Ali Ahmed received his MSc in Electronics & Communication Engineering, Sudan University of Science & technology in 2012 and BSc in 2008. He worked as a senior engineer in Sudan Sea Port Corporation for five years as a team leader of new projects in wireless networks, including Tetra system, Wi-Fi, WI-Max, and CCTV. In addition, he also worked as a senior lecturer and Head of Electrical and Electronics Engineering Department in the Red Sea University. Currently he is the head of marine systems department in Sudan Marine Industries. Elmustafa has published more than 45 research papers and book chapters in wireless communications systems and networking in peer-reviewed academic international journals. His areas of research interest include routing protocols, computer and wireless networks, Internet-of-Things (IoT) Applications, Underwater Communications, and Artificial Intelligence (AI) in wireless networks. He is a member of IEEE Communication Society (ComSoc), International Association of Engineers (IAENG), Six Sigma Yellow Belt (SSYB), Scrum Fundamentals certified (SFC), and Sudanese Engineering Council (SEC).

Saad Aslam received his PhD degree in Electrical and Electronic Engineering from Massey University, New Zealand. He is currently a Senior Lecturer at the School of Engineering and Technology, Sunway University, Malaysia. He has more than 12 years of experience in academia blended with industry exposure as well. His current research involves exploiting machine learning for optimizing wireless networks. Generally, his research focuses on D2D communication, clustering algorithms, distributed systems, and game theory optimization.

Mohammed Abdul Azim received his PhD in Electrical and Information Engineering, University of Sydney, Australia, 2007. Currently he is a post-doctoral fellow at Masdar Institute of Science and Technology (MIST), Abu Dhabi, UAE. He is researching on wireless sensor network localization and cooperative communication. He was a senior researcher at Malaysian Institute of Microelectronic Systems (MIMOS) Berhad, focusing on wireless network protocol especially on a MIMOS product that combines Wi-MAX and Wi-Fi. Mohammed has published many research papers, books, and book chapters on wireless communications and networking in peer-reviewed academic journals and conferences.

Linga Reddy Cenkeramaddi received his master's degree in electrical engineering from the Indian Institute of Technology, Delhi (IIT Delhi), India, in 2004, and the PhD degree in Electrical Engineering from the Norwegian University of Science and Technology (NTNU), Trondheim, Norway, in 2011. He worked for Texas Instruments in mixed signal circuit design before joining the PhD program at NTNU. After finishing his PhD, he worked in radiation imaging for an atmosphere–space interaction monitor (ASIM mission to the International Space Station) at the University of Bergen, Norway from 2010 to 2012. At present, he is the group leader of the autonomous and cyber-physical systems (ACPS) research group and is working as a professor at the University of Agder, Campus Grimstad, Norway. He has co-authored more than 100 research publications that have been published in prestigious international journals and standard conferences. His main scientific interests are in cyber-physical systems, autonomous systems, and wireless embedded systems. He is the principal and co-principal investigator of many research grants from the Norwegian Research Council. He is a member of the editorial boards of

various international journals, as well as the technical program committees of several IEEE conferences.

Soumaya Cherkaoui (Senior Member, IEEE) is a full professor at the Department of Computer and Software Engineering at Polytechnique Montreal. Before joining academia as a professor in 1999, she worked for industry as a project leader on projects targeted at the Aerospace Industry. She leads a research group which conducts research funded by both government and industry. Her research interests are in wireless networks. Particularly, she works on the next generation 5G/6G networks, edge computing/network intelligence, and communication networks for verticals such as connected and autonomous cars, IoT, and industrial IoT. Her work resulted in technology transfer to companies and to patented technology. She has published more than 200 research papers in reputed journals and conferences. She has been on the editorial board of several IEEE journals. Her work was awarded with recognitions including several best paper awards at IEEE conferences. She was designated by the IEEE Communications Society as Chair of the IoT and Ad Hoc and Sensor Networks Technical Committee in 2020. She is a professional engineer in Canada and an IEEE ComSoc distinguished lecturer.

Mona Bakri Hassan received her MSc in electronics engineering and telecommunication in 2020 and BSc (Honors) in electronics engineering and telecommunication in 2013. She has completed Cisco Certified Network Professional (CCNP) routing and switching course and passed the Cisco Certified Network Administrator (CCNA) routing and switching. She has gained experience in network operations center as a network engineer in back office department in Sudan Telecom Company (SudaTel) and in vision valley as well. She has also worked as a teacher assistant in Sudan University of Science and Technology which has given her a great opportunity to gain teaching experience. She published book chapters and papers on wireless communications and networking in peer-reviewed academic international journals. Her areas of research interest include cellular LPWAN, IoT, and wireless communication.

Sindhusha Jeeru is currently working as a visiting researcher at the University of Agder, Norway. She received her Master of Science degree in Telecommunication from the University of Jaen, Spain, in 2020, and BE in Electronics and Communication Engineering from Andhra University in 2018. Her research interests mainly include Wireless Communication, IoT, and Signal Processing.

Lyes Khoukhi (Senior Member, IEEE) received the PhD degree in electrical and computer engineering from the University of Sherbrooke, Canada, in 2006. From 2007 to 2008, he was a post-doctoral researcher with the Department of Computer Science and Operations Research, University of Montreal. Since September 2020, he has been a full professor with the ENSICAEN, GREYC Laboratory, Caen, France. Previously, he was associate professor with the University of Technology of Troyes. His current research interests include network security, blockchain, and advanced network.

Zoubeir Mlika (Member, IEEE) received his BE degree from the Higher School of Communication of Tunis, Tunisia, in 2011 and the MSc and PhD degrees from the University of Quebec at Montreal (UQAM), Canada, in 2014 and 2019, respectively. He was a Visiting Research Intern with the Department of Computer Science, University of Moncton, Canada, between March 2018 and August 2018. In September 2019, he joined the Department of Electrical and Computer Engineering, University of Sherbrooke, Canada, as a post-doctoral fellow. His current research interests include vehicular communications, resource allocation, reinforcement learning, machine learning, design and analysis of algorithms, computational complexity, and quantum machine learning.

Hajar Moudoud (Member, IEEE) received her BE degree in software engineering from the Mohammadia School of Engineers, Rabat, Morocco, the first PhD degree in computer engineering from the University of Sherbrooke, Canada, and the second PhD degree in computer engineering from the University of Technology of Troyes, France, in 2018 and 2022, respectively. Her research interests include security of Internet-of-Things, applied machine/deep learning for intrusion detection system, and integration with blockchain.

Muhammad Nadeem is a full-time assistant professor in the School of Engineering and Technology at the American University of the Middle East. He earned a bachelor's degree in Electrical Engineering from the University of Engineering and Technology Lahore, and a Master of Science degree from the Royal Institute of Technology, Sweden. He received his PhD degree in Electronics Engineering from the University of Auckland, New Zealand. His research interests include embedded systems, IoT, augmented reality, and machine learning.

Houshyar Honar Pajooh received his PhD degree in electronics and computer engineering from Massey University, Auckland, New Zealand. He has more than 15 years of experience working in ITC and telecommunications, and since 2015, he has been involved in cybersecurity and computer science research with a strong focus on applied cryptography, network security, and distributed computing.

The goal of his research is to develop smart systems and methodologies, with a focus on practical implementation and real-world feasibility. His research is multidisciplinary, and his current fields of interest include applied cryptography, network security, machine learning, IoT, information security, and applied blockchain and cryptography.

Mamoon Mohammed Ali Saeed is a lecturer at the Department of Communication and Electronics Engineering, UMS University, Yemen. He received his bachelor's in communication and Electronics Engineering from Sana'a University, Yemen 2005, and MSc degree in Computer Networks and Information Security in Yemen Academy for Graduate Studies, Yemen, in 2013. He received his PhD degree from Alzaiem Alazhari University, Faculty of Engineering, and Electrical Engineering Department, Sudan 2022. His research areas include information security, communication security, and network security.

Rashid A. Saeed received his PhD in Communications and Network Engineering, Universiti Putra Malaysia (UPM). Currently he is a professor in the Department of Computer Engineering, Taif University. He is also working in Electronics Department, Sudan University of Science and Technology (SUST). He was a senior researcher in Telekom Malaysia™ Research and Development (TMRND) and MIMOS. Rashid published more than 150 research papers, books and book chapters on wireless communications and networking in peer-reviewed academic journals and conferences. His areas of research interest include computer network, cognitive computing, computer engineering, wireless broadband, and WiMAX Femtocell. He has been successfully awarded three US patents in these areas. He supervised more 50 MSc/PhD students. Rashid is a senior member IEEE, Member in IEM (I.E.M), SigmaXi, and SEC.

Afaf Taik (Graduate Student Member, IEEE) received her BE degree in software engineering from the Ecole Nationale Supérieure d'Informatique et d'Analyse des Systèmes, Rabat, Morocco, in 2018, and a DESS in applied sciences from the Université de Sherbrooke, Canada in 2018. She is currently pursuing her PhD degree in Electrical and Computer and Electrical Engineering, Université de Sherbrooke, Canada. Her research interests include edge computing, machine learning, and the development of statistical and mathematical methods for distributed systems in wireless networks.

Anurag Thantharate is a member of the IEEE who received his MS degree in electrical engineering from the University of Missouri, Kansas City, US, in 2012 and is currently pursuing his PhD in computer networking and communication systems at the university. He is also a senior manager of technology at Palo Alto Networks, Santa Clara, California, where he works on 5G, Network Slicing, Multi-Access Edge Computing and Cloud Security and has previously worked with Samsung Electronics, T-Mobile, Sprint, Qualcomm, and AT&T Mobility over the past 11 years on a broader aspect of wireless network and technologies areas. His research interests include applications of machine learning in wireless communication, network slicing, and autonomous network management toward zero touch approach.

Sreenivasa Reddy Yeduri received his BE degree in electronics and communication engineering from Andhra University, Visakhapatnam, India, in 2013; the MTech degree from the Indian Institute of Information Technology, Gwalior, India, in 2016; and the PhD degree in Electronics and Communication Engineering, National Institute of Technology, Goa-India, in 2021. He is currently working as a post-doctoral research fellow with autonomous and cyber-physical systems (ACPS) research group, Department of ICT, at University of Agder, Grimstad, Norway. His research interests are machine-type communications, Internet-of-Things, LTE MAC, 5G MAC, optimization in communication, wireless networks, power line communications, visible light communications, hybrid communication systems, spectrum cartography, spectrum sensing, UAV-assisted wireless networks, V2X communication, V2V communication, wireless sensor networks, mmWave RADAR, and aerial vehicle traffic management.

Chapter 1

Green machine learning for cellular networks

Saad Aslam and Houshyar Honar Pajooh
Sunway University, Kuala Lumpur, Malaysia

Muhammad Nadeem
American University of the Middle East, Egaila, Kuwait

Fakhrul Alam
Massey University, Auckland, New Zealand

1.1 INTRODUCTION

Global energy consumption continues to grow. One of the contributing factors is the increased computational capability of devices. Many disciplines, e.g., engineering, computer science, economics, management, business, are utilizing machine learning (ML) for finding optimal solutions for their respective applications.

As is the case with other systems, cellular networks need to reduce their energy consumption. Energy-efficient cellular networks will help cope with climate change and protect the environment. Providing energy-efficient solutions for cellular networks is becoming more important with every coming generation of technology. The reason being the number of devices connected to cellular networks is increasing exponentially (see Figure 1.1). This growth in the number of devices increases the overall energy consumption attributed to devices as well. Network energy consumptions are likely to increase manifolds in the upcoming years [1]. This massive increase in the volume of network traffic brings a challenging scenario. According to recent reports [2], CT infrastructure consumes 3% of the worldwide energy and thereby causes 2% of the worldwide CO_2 emissions.

It should be noted that a major portion of traffic belongs to cellular and mobile networks [1]. Therefore, optimizing the energy consumption of cellular networks is of utmost importance. It has twofold advantages: one is minimizing the environmental impact, and the other is minimizing the network operational costs.

The beyond 5G (B5G) and 6G cellular networks are expected to address complex network issues. This increase in network complexity is due to the emergence of new applications and use cases, physical layer design, and the requirements of today's data-hungry users [3]. New standards are being

DOI: 10.1201/9781003230427-1

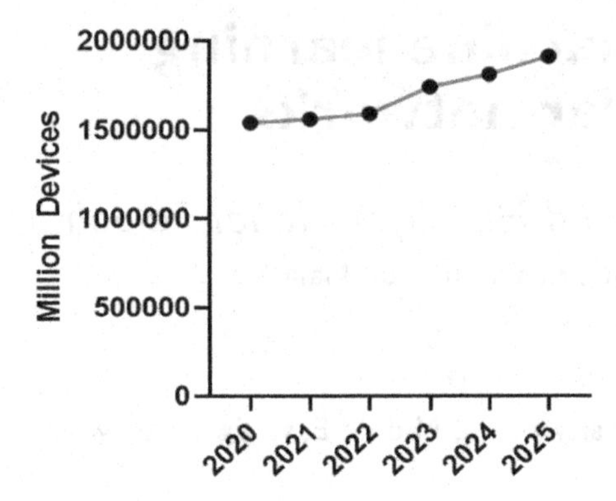

Figure 1.1 The number of devices connected to cellular networks is expected to increase at a significantly high rate.

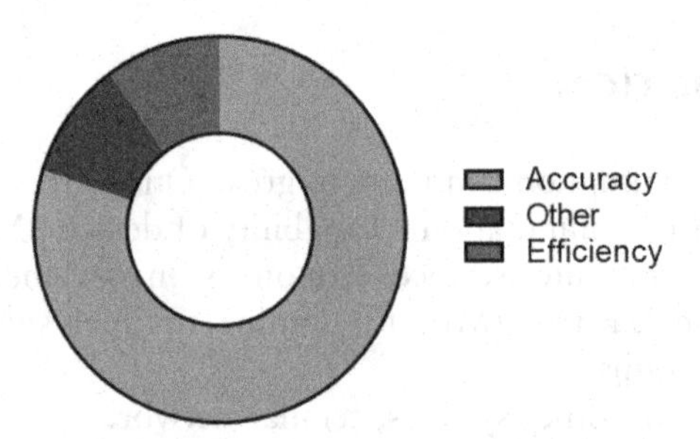

Figure 1.2 Data is taken from 30 research articles (journals and conferences related to cellular networks only) published during 2019–2021. All these research articles consider ML for cellular networks. It can be seen that accuracy dominates the performance metric utilized for evaluating the ML algorithms.

introduced frequently to address these issues. ML has seen remarkable progress in finding optimal solutions to various problems from the physical layer to the application layer of the cellular network [4]. ML can provide simplified solutions to improve the performance of the network and is therefore considered to be an integral part of B5G and 6G network architecture.

The effectiveness of the ML-based solutions is typically evaluated by the accuracy of the algorithm (see Figure 1.2 for more details). Energy consumption or cost of the ML implementation is often not considered. This emphasis on a single performance evaluation metric ignores other important criteria such as economic, environmental, and social costs of reaching the required accuracy [5]. Therefore, this chapter promotes the investigation of methodologies, mechanisms, and algorithms that can produce environment-friendly solutions. It is recommended to make efficiency an important evaluation criterion.

Several factors can increase the computational complexity of ML algorithms. All these factors also provide opportunities to develop efficient solutions. To realize this, it is important to investigate the cost and computational complexity of developing, training, and running the models. Most of the recent studies only consider training cost. Further, it is important to consider the size of the training set as it would be interesting to investigate the impact of reducing the training set on the accuracy of the trained models.

To provide sustainable ML solutions for cellular networks, it is important to explore the concepts of green machine learning (GML). Therefore, in this chapter, we highlight the importance of introducing GML in the computational world for reducing adverse impacts on the environment. We discuss various performance parameters that are often not considered for evaluating the performance of ML algorithms. This chapter will also present techniques to decrease the computational complexity of the ML algorithm. Finally, the challenges of bringing GML to the computational world will be discussed in detail.

The organization of this chapter is as follows. Section 1.2 discusses the challenges of GML, whereas Section 1.3 describes the conventional ML models with emphasis on DL and provides the reasoning about why DL is a widely used and implemented technique. Section 1.4 presents initial studies on the topic of GML. The next section is dedicated to understanding the difference between the performance evaluation criteria of conventional ML and GML. Section 1.6 provides a detailed overview of the research directions/possible steps for implementing GML. The last section concludes this chapter. The list of acronyms can be found in Table 1.1.

Table 1.1 List of acronyms

Adversarial Neural Network	ANN
Beyond Fifth Generation	B5G
Carbon Dioxide	CO_2
Communication Technologies	CT
Convolutional Neural Network	CNN
Deep Learning	DL
Deep Neural Network	DNN
Floating-point Operations	FPO
Graphics Processing Unit	GPU
Green Machine Learning	GML
Green Artificial Intelligence	Green AI
Machine Learning	ML
Natural Language Processing	NLP
Probabilistic Programming	ProbP
Recurrent Neural Network	RNN
Sixth Generation	6G

1.2 THE GREEN ML CHALLENGE

ML is a ubiquitous phenomenon. During the last few years, researchers have proposed many ML-based approaches for optimizing cellular networks. Recent literature suggests that most of these approaches are deep learning (DL) based. This trend is not limited to cellular networks; rather, DL-based approaches have brought benefits to other disciplines, such as computer vision, natural language processing, and non-intrusive electrical load monitoring. However, these advancements come at a cost of huge computational resources. The drive for ever-increasing accuracy has made the energy consumption of devices increase as well, resulting in a much larger carbon footprint. It should be noted that most of the recent ML models consist of complicated architectures that consume a vast amount of energy. Since cellular networks are expected to extensively employ ML [1, 3, 4], it is imperative to pay attention to energy usage and carbon emission during model training and inference. We believe that more attention and effort to go into lightweight and energy-efficient algorithms is necessary.

Moreover, this discussion on the lack of focus on energy consumption is not only relevant to certain types of ML algorithms or certain applications of cellular networks. Without new approaches, ML's energy impact could quickly become unsustainable.

To begin the investigation on energy consumption, we need a good understanding of the relationship between the approaches used to train a model and the energy they require.

1.3 THE CONVENTIONAL MACHINE LEARNING MODELS: HEAVY RELIANCE ON DEEP LEARNING

Machine learning models are resource-hungry (in terms of GPU, energy consumption, and computational complexity) algorithms, which is an even bigger drawback for DL models. Research presented in [6, 7] suggests substantial energy is consumed while training neural networks. The initial works related to ML and AI started with shallow models, e.g., Conditional fields (Ghosh et al. [8]) which require considerably less computational energy as compared to the recent DL models. It shows that we are moving in the opposite direction to GML.

Easy availability of powerful GPUs is an enabling factor for the growth of complex ML algorithms. Moreover, the emergence of Convolutional Neural Network (CNN), Recurrent Neural Network (RNN), Adversarial Neural Network (ANN), etc. has influenced researchers working with ML projects to actively use these powerful tools. These models provide many advantages but at the cost of a tremendous increase in the consumption of energy [6, 7].

The research problems of cellular networks that were previously investigated with conventional techniques are now being reinvestigated with DL.

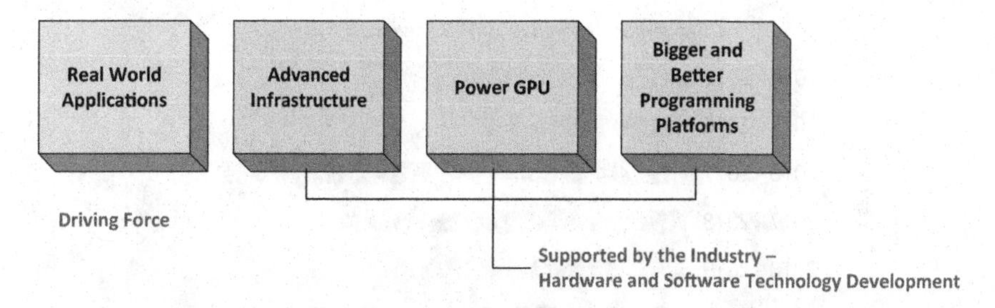

Figure 1.3 The heavy reliance on deep learning algorithms stems from the fact that it targets numerous real-life applications and it has been well supported by industrial advancements (hardware and software).

The reason is the desire to improve accuracy. Even if we think beyond cellular networks, one of the advantages of using DL is its applicability to real-world applications. Figure 1.3 represents the appeal of using DL.

Moreover, if an application targets real-world scenarios, the industry is always willing to develop and expand complementary technology around that. Therefore, the advancement in infrastructures, hardware, software, etc. has enabled researchers to develop even deeper and more powerful algorithms that achieve better results in terms of accuracy [6].

Seventy ML-related research papers, published between 2014 and 2021, that specifically employ deep learning are chosen randomly. These papers were selected from multiple databases/publishers (Scopus, PubMed, Web of Science, Science Direct, IEEE Xplore, etc.). Figure 1.4 represents the percentage increase in the usage of deep learning algorithms from 2014 to 2021. The y-axis represents the percentage increase in the use of DL algorithms in a particular year range. It should be noted that these percentages should be taken as an approximate estimate and not an exact value. To provide more insights, we have presented two observations, Figure 1.4(a) represents research articles published in domains other than cellular networks, and Figure 1.4(b) shows cellular networks-related research.

A significant increase in the usage of DL can be seen over the past many years. However, this trend is even stronger for research focusing on cellular networks. It shows the need to attract researchers working in this domain toward considering energy efficiency while developing ML-related solutions.

1.4 TOWARD GREEN MACHINE LEARNING: SOME RELEVANT STUDIES

Though researchers have started investigating GML protocols and techniques for sustainable solutions, significant research on this area is still lacking and many questions need to be answered. However, a number of pioneering studies can be found in the literature, and they are briefly presented next.

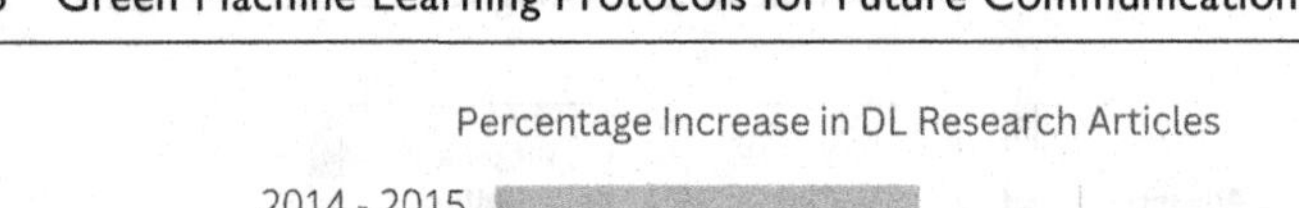

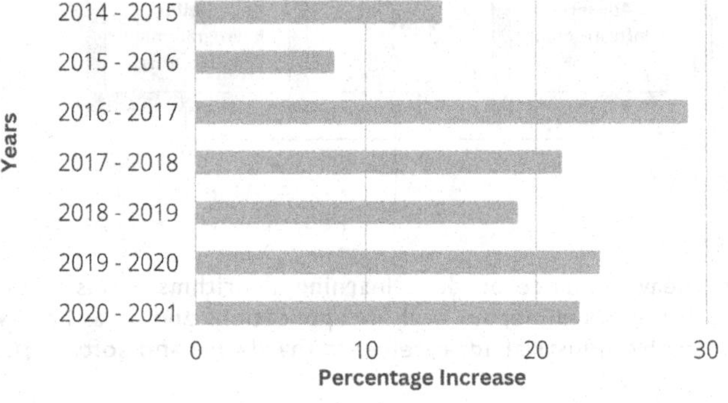

(a)

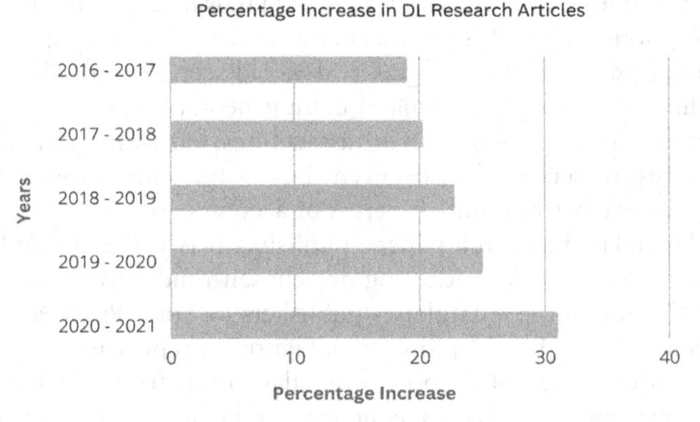

(b)

Figure 1.4 Most of the ML-related research works are extensively using DL. The percentage increase in the usage of DL is presented. A significant increase is seen over the past seven years. (a) Research published in various domains – excluding cellular networks. (b) Research published in cellular networks domain.

One of the earliest works that discusses the computational requirements of ML implementations has been presented by Li et al. [9]. This study presents the details on the energy consumption required for image classification using CNN. Different network models were presented to compare the energy consumption for the same application while both CPU and GPU are utilized to train a model. A typical CNN network can consume around 87% of the total energy consumption of the computing device and contribute significantly to the carbon footprint. It was reported that even per iteration energy consumption goes on increasing as we continue to optimize the learning model.

Another research conducted by Canziani et al. [10] sheds light on the GPU power consumption for the image classification model. The model size was varied to provide a comparative study as well. Similar trends, as reported in [9], were observed in this work as well. It should be noted that both studies considered similar applications and the same ML algorithms. Further, most of the deeper and more complex models (such as NLP and RNN) that we widely use today were not evaluated in these studies, and even then, energy consumptions were found to be quite substantial.

One of the most comprehensive studies that evaluated the impact of training deep learning models on the carbon footprint is presented in [6]. This paper provided an early signal about the need to move from the red zone of energy consumption to the GML paradigm. The authors estimated the CO_2 emissions from training a single Natural Language Processing (NLP) model. The emissions were significantly larger than those of a vehicle spanning its lifetime. These startling results have persuaded some researchers to think "Green" about ML-based implementations.

Another interesting study was conducted in 2020 which introduced the "Green AI" concept. It emphasized using computational cost as a key performance indicator for ML algorithms [11]. This work also provided an estimate of the increase in the computational costs that have transpired in recent years. Over the past seven years, computational complexity has gone up by 300,000 times, therefore contributing to a larger footprint.

The study presented in [12] has identified a variety of factors that increase the cost of training ML algorithms. This work also details the risk associated with the trend of developing deep learning models with ever-increasing data sets. These risk factors include financial (escalating demand for bigger and better computational systems), environmental (larger carbon footprints), and less research opportunity (less focus on Green Computations).

Another study [13] estimates the energy usage for training computationally complex ML algorithms and determines its impact on the environment. This work presents an interesting finding about the factors that can help reduce energy use. Choice of DNN, geographical location where ML algorithms are trained, and the type of data center infrastructure all can contribute toward a greener computation [13]. According to the findings of this work, "large but sparsely activated DNNs" should be used. Moreover, when and where ML algorithms are trained should be considered as well. The type of data center infrastructure, i.e., whether it is cloud-based or does it use ML accelerators, can significantly impact the carbon footprint. In a nutshell, smarter choices around these factors can reduce the carbon footprint up to ~100–1000X [13].

1.5 KEY PERFORMANCE INDICATORS: CONVENTIONAL VERSUS GREEN COMPUTATIONS

The conventional ML approaches aim to improve the accuracy of the training algorithms. This emphasis on accuracy disregards important factors like model complexity, size of the training dataset, number of hyperparameters, computational power, and energy usage. Interestingly, the recent literature suggests that in many reported works, all the above-mentioned factors have shown a strong (positive) correlation with accuracy, i.e., increasing the dataset, computational power, and model complexity increases the accuracy as well. We randomly picked 75 research articles, from several databases/publishers (Scopus, PubMed, Web of Science, Science Direct, IEEE Xplore, etc.) within the relevant ML domain, and gathered the performance criteria of these papers. The result is demonstrated in Figure 1.5. It should be noted that while Figure 1.2 provides insights relevant to cellular networks only, Figure 1.5 considers various research fields, demonstrating a trend not limited to cellular networks.

The result shows a consistent trend; accuracy is the most used metric for evaluation of the ML-based solutions. Moreover, it is worth noting that the efficiency is being neglected as an important metric of evaluation.

The other important observation from this collection of papers is that the researchers are continuously looking for increasingly large and complex models (in terms of layers of the deep network, hyperparameters, activation functions, etc.) and therefore, the demand for powerful computational machines is rising as well. The problem with such large models is that the cost of processing even simple datasets is increasing. More complex and costly objective

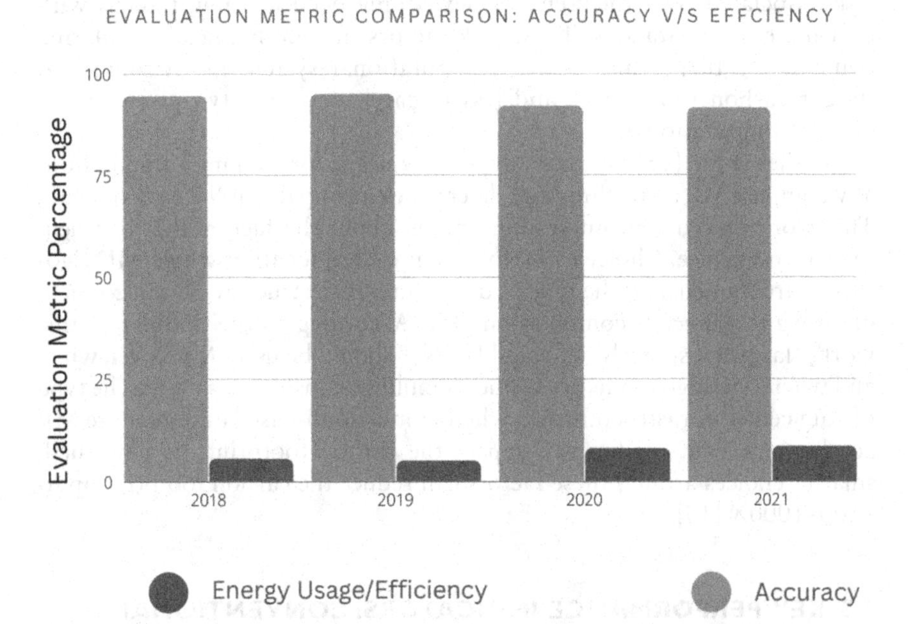

Figure 1.5 Over the four-year period of 2018 to 2021, accuracy is one of the most commonly used performance metrics whereas only a handful of research works targeted energy efficiency.

functions are being used that are consuming more power. It shows that the financial and environmental effects of these choices are not being considered.

To move toward GML, contrary to recent research trends, we believe that alternative performance parameters should be targeted. Here we present a discussion around these parameters.

A. Energy Consumption

Energy consumption is an important parameter that should be considered while evaluating an ML algorithm [11]. The focus on evaluating the amount of energy required to produce the intended result will bring efficiency into the picture. Moreover, there should be a detailed overview of energy consumption at each stage of training an algorithm, i.e., energy consumed for training a model, testing a model, data augmentation (if required), and hyperparameters tuning as well. This decomposition of overall energy consumption gives researchers a starting point to optimize energy usage or reduce energy usage wherever possible throughout the training/experiment.

B. Carbon Emission

As we promote moving toward GML, evaluation of carbon emissions is a must. It is that one parameter that we want to reduce as much as possible. It is directly interfering with our environment. As discussed in [11], while it might not be straightforward to determine the CO_2 emissions for a particular ML model, an estimate can be made based on the electricity infrastructure.

C. Computing Parameters

A few computing parameters that might be useful in evaluating an algorithm are: choice of model (linear, non-linear, DL, or Non-DL based), total time needed to produce a result from a model, and the number of hyperparameters utilized. Some of the important works in this domain [6, 11] also utilize Floating-point Operations (FPOs) as an important parameter for understanding the computational cost of an algorithm. FPO is directly tied to the energy consumption of a particular model. However, it should be noted that FPO is closely dependent on the hardware used for executing a particular task.

D. Hardware Performance

Energy consumption is related to the processing hardware as well [11]. Therefore, it is important to report hardware-specific parameters, e.g., average CPU/GPU usage for training the same dataset. This will help in evaluating the performance of a given algorithm on particular hardware. This information will also provide an opportunity for the researchers to identify hardware requirements for a given task, utilizing optimal energy, as compared to opting for other hardware consuming more energy for processing the same dataset.

1.6 TOWARD GREENER ML

A. **Information-Centric Networking for Cellular Networks**
Obtaining an accurate machine learning model can be computationally complex for both the devices/mobile users and the network. One of the strategies that can be used to reduce the computational complexity and energy consumption of devices is to utilize information-centric networking [14]. It has been reported that data gathering and storage consume significant energy in cellular networks. Information-centric networking exploits the availability of data from various sources, gathering small portions of large datasets [14, 15]. These sources can be used to formulate a multi-source optimization problem to reduce the energy consumption of the devices and the network involved with the ML algorithm.

B. **Energy-efficient IT Infrastructure for Cellular Networks**
As ML becomes deeply involved with networks, dedicated machine learning servers and data centers are now being introduced [16]. Therefore, it is important to have the right infrastructure for energy-efficient processing. One of the infrastructure-related decisions is to have decentralized processing centers, the so-called hybrid processing environment, where cloud, co-located, and remote processing facilities, and dedicated data centers can all work together [16]. It is important since significant power is utilized to run data centers and their centralized processing and management schemes can actually lead to increased power utilization and thus CO_2 emissions as well. Banking on cloud computing and edge computing technologies, both hardware- and software-level changes are required to bring data centers closer to greener processing.

C. **Probabilistic Programming**
Probabilistic programming (ProbP) can facilitate move toward Greener ML [17]. As opposed to millions of data points required to train an ML algorithm, ProbP can work on considerably smaller datasets and express the output "smartly." ProbP understands the outcomes of previously generated models and provides an opportunity to statically transfer this learning to the new model.

D. **Transfer Learning and Reproducibility**
ML has been widely used in all domains; numerous trained algorithms are readily available. Therefore, a fundamental question here would be whether we need to produce a new model in all cases. It is understood that available models may not suit every application or every scenario. However, there can be many potential overlaps [18].

One of the approaches that could be utilized is transfer learning [18] where an already trained model is re-structured for a different task. Moreover, it is also important to share information about the trained model as much as possible to help subsequent studies make direct use of already available information and help in reproducing the already-conducted research. It might open an opportunity for saving/reducing computational costs.

E. **Improved Computing Hardware**
It is vital to invest more in the development of efficient hardware. Better hardware can utilize resources (power, energy, electricity usage, etc.) more efficiently. Secondly, efficient hardware might bring repeatability to the energy usage; i.e., similar experiments ran at different times and different geographical locations bring out the same results at the same energy cost.
Algorithm complexity influences the choice of hardware as well. The requirement for a high-spec computational machine increases with the increase in the parameters of an algorithm. High-performance and energy-efficient hardware can speed up the learning process and complete the task in a shorter time with less energy and electricity usage, thus reducing carbon footprint.

F. **Non-DL Models are effective as well**
Deep learning has indeed provided excellent results in many disciplines. However, DL is not the solution for every ML problem. It has been seen in many cases that researchers have preferred DL for implementation even with marginal benefits in accuracy. Therefore, we are over-reliant on a complex and computationally hungry algorithm (please see Figure 1.4).

G. **Energy-Efficient Training**
There are many stages to creating and implementing ML models [19]. Here we focus on the training stage. Training starts from hyperparameters optimization. A lot of computations are required to reach the optimal solution. Since appropriate initialization of hyperparameters can help the algorithm to converge timely, this stage directly impacts the energy efficiency of the algorithm. There are numerous initialization strategies (e.g., Random Initialization) that should be carefully selected befitting to the chosen algorithm and application. Moreover, other-than-conventional hyperparameters optimization techniques that have been proposed in recent years [18, 19] should be investigated as well.

Moreover, on a separate note, the literature presents a noteworthy observation. Sometimes increasing 1% of accuracy can lead to an exponential rise in energy consumption [17]. It means these parameters exhibit non-linear behavior, and therefore, researchers should decide if it is worth increasing accuracy by a few percent while energy consumption explodes. Along with this, it is also important to have answers to the following questions: How much accuracy is enough for a particular application, and how much data would be required to achieve that accuracy? One of the approaches that can help answer these questions could be active learning [18]. This field of study considers using fewer samples to achieve the intended results. The idea is to divide a big dataset into several pools and find the most useful data pool to train the model. It might help in eliminating the pool/data samples which are not helpful in learning.

1.7 CONCLUSION

The cellular network is one of the many fields that have benefited from the advances in ML. These advances have encouraged many researchers to develop complex, data-hungry, and energy-consuming algorithms. A significant number of research articles related to cellular networks are using complex and deep learning strategies. Unfortunately, most researchers have overlooked its energy impact. Therefore, this is a good time to make a move toward GML. It should be noted here that this research is not against conventional machine learning algorithms. It is understood that many fields have progressed extraordinarily owing to the development of ML algorithms. However, reducing carbon footprint matters, and one of the potential steps can be to invest more in GML.

In this chapter, we have identified the challenges in moving toward GML. It is observed that current research trends are substantially dependent on deep learning and the focus of most of the articles is to improve the accuracy of the trained model. Therefore, currently, we are in the red zone of energy consumption. Further, we have introduced various performance evaluation metrics that are required for energy-efficient machine learning models. It is emphasized that energy efficiency, parameters related to hardware performance, computing-specific parameters, and, most importantly, carbon emissions should be noted for the performance evaluation of a particular ML model. Lastly, detailed discussion and recommendations have been provided on the possible steps that need to be taken for GML protocols including, cellular-specific technology improvements, new and improved probabilistic programming techniques, efficient training practices, and hardware improvements as well.

In the future, it would be interesting to investigate the impact on accuracy as we introduce energy-efficient ML algorithms. A comparative study of conventional and energy-efficient algorithms with respect to accuracy would be beneficial. A trade-off can be presented about making an algorithm greener as compared to making the same more accurate. Moreover, efficient hyperparameter optimization techniques should be studied. It is mentioned in the previous section that distributed learning is important for GML; therefore, it is important to explore cellular network architectures that can accommodate such changes.

REFERENCES

[1] Gür, Gürkan, and Fatih Alagöz. "Green wireless communications via cognitive dimension: an overview." *IEEE Network* 25.2 (2011): 50–56.

[2] The Climate Group Report, "SMART2020: Enabling the Low Carbon Economy in the Information Age," http://www.theclimategroup.org, September 2015.

[3] Challita, Ursula, Henrik Ryden, and Hugo Tullberg. "When machine learning meets wireless cellular networks: Deployment, challenges, and applications." *IEEE Communications Magazine* 58.6 (2020): 12–18.

[4] Aslam, S., F. Alam, S. F. Hasan, and M. A. Rashid, "A machine learning approach to enhance the performance of D2D-enabled clustered networks." *IEEE Access* 9 (2021): 16114–16132. https://doi.org/10.1109/ACCESS.2021.3053045

[5] Candelieri, Antonio, Riccardo Perego, and Francesco Archetti. "Green machine learning via augmented Gaussian processes and multi-information source optimization." *Soft Computing*, 25 (2021): 1–13.

[6] Strubell, Emma, Ananya Ganesh, and Andrew McCallum. "Energy and policy considerations for deep learning in NLP." arXiv preprint arXiv:1906.02243 (2019).

[7] Hao, Karen. "Training a single AI model can emit as much carbon as five cars in their lifetimes." *MIT Technology Review* (2019).

[8] Ghosh, S., R. Johansson, G. Riccardi, and S. Tonelli (2011, November). Shallow discourse parsing with conditional random fields. In *Proceedings of 5th International Joint Conference on Natural Language Processing* (pp. 1071–1079).

[9] Li, Da, Xinbo Chen, Michela Becchi, and Ziliang Zong (2016). Evaluating the energy efficiency of deep convolutional neural networks on cpus and gpus. *2016 IEEE International Conferences on Big Data and Cloud Computing (BDCloud), Social Computing and Networking (SocialCom), Sustainable Computing and Communications (SustainCom) (BDCloud-SocialCom-SustainCom)*, pp. 477–484.

[10] Canziani, Alfredo, Adam Paszke, and Eugenio Culurciello. 2016. An analysis of deep neural network models for practical applications.

[11] Schwartz, Roy, Jesse Dodge, Noah A. Smith, and Oren Etzioni. "Green AI." *Communications of the ACM* 63.12 (December 2020), 54–63. https://doi.org/10.1145/3381831

[12] Bender, Emily M., et al. (n.d.). "On the Dangers of Stochastic Parrots: Can Language Models Be Too Big?" *Proceedings of the 2021 ACM Conference on Fairness, Accountability, and Transparency.*

[13] Patterson, David, et al. (2021). "Carbon emissions and large neural network training." arXiv preprint arXiv:2104.10350.

[14] Yang, Y. and T. Song, "Energy-Efficient Cooperative Caching for Information-Centric Wireless Sensor Networking," *IEEE Internet of Things Journal* 9.2 (15 January, 2022), 846–857. https://doi.org/10.1109/JIOT.2021.3088847

[15] Aslam, S., F. Alam, H.H. Pajooh, M.A. Rashid, and H.M. Asif (2021). Machine learning applications for heterogeneous networks. In Al-Turjman, F. (eds) *Real-Time Intelligence for Heterogeneous Networks*. Springer, Cham. https://doi.org/10.1007/978-3-030-75614-7_1

[16] Katal, Avita, Susheela Dahiya, and Tanupriya Choudhury. "Energy efficiency in cloud computing data centers: a survey on software technologies." *Cluster Computing* (2022), 1–31.
[17] Belcher, Lewis, Johan Gudmundsson, and Michael Green. "Borch: A Deep Universal Probabilistic Programming Language." arXiv preprint arXiv:2209. 06168 (2022).
[18] Xu, J., W. Zhou, Z. Fu, H. Zhou, and L. Li (2021). A survey on green deep learning. arXiv preprint arXiv:2111.05193.
[19] Sharma, Vibhu and Vikrant Kaulgud, Accenture website, accessed 26 September 2022. https://www.accenture.com/us-en/blogs/technology-innovation/sharma-kaulgud-developing-energy-efficient-machine-learning

Chapter 2

Green machine learning protocols for cellular communication

Mamoon Mohammed Ali Saeed
University of Modern Sciences (UMS), Sana'a, Yemen

Elmustafa Sayed Ali Ahmed
Red Sea University, Port Sudan, Sudan
Sudan University of Science and Technology, Khartoum, Sudan

Rashid A. Saeed
Taif University, Makkah, Saudi Arabia
Sudan University of Science and Technology, Khartoum, Sudan

Mohammed Abdul Azim
Asian University for Women, Chattogram, Bangladesh

2.1 INTRODUCTION

Technology for communication is essential in today's environment. Most activities in the actual world are managed by machines and computers. Information and communications technology developments have made all of this feasible. Due to the incredible innovation and progress in electrical and communication technologies, the data rate in cellular communication is rising quickly. Researchers worry that the quick advancement of these technologies could cause a huge increase in data traffic. The 4G and 5G cellular communication generations now in use are becoming more and more congested, with issues like capacity, bandwidth scarcity, interference, and slower data speeds. Higher data rates and fast internet will be essential for these connected devices [1–3].

To support all communications devices, a large amount of transmit power would be required, resulting in significant energy consumption. Greenhouse gases (GHG) emissions would result as a result. In the future, the amount of carbon dioxide released by these interconnected devices will rise by millions of tons. To make the future safer, CO_2 emissions, one of the factors contributing to global warming, should be taken seriously. The Federal Communications Commission (FCC) has established a threshold for surface area radiations and restricted the transmit power for portable devices to safeguard human

DOI: 10.1201/9781003230427-2

users from electromagnetic exposure. How much radioactive material is absorbed by human tissue is quantified using the Specific Absorption Rate (SAR) statistic. A system with a higher SAR value will be detrimental to both human health and the ecosystem [4].

Energy efficiency is essential to ensuring that wireless devices have longer battery lives. The battery in these devices needs to be more effective to run for longer periods of time as the number of mobile phone users rises. This method is being used in several studies to advance battery technology for electric vehicles and mobile phones. But compared to its application, this field of research is developing and progressing far more slowly. As a result, making reusable equipment and using energy-efficient communication are beneficial expenditures [5].

Utilizing energy-efficient equipment and managing cellular network power usage are two key components of green communications. One method for reducing power consumption and improving the energy efficiency of networks is resource allocation, but it has a number of drawbacks of its own. The network design can further reduce power consumption. Green machine learning is therefore urgently needed, and as information and communications technology advance, it is becoming a more interesting subject [5].

The development of modern society has increased interest in green technology. Likewise, as communication technology develops, business and researchers work to make these communications as environmentally friendly as they can be. The next step in satisfying consumer demands is the development of 5G cellular technology. In this chapter, various green communication technologies are examined, along with certain issues [6].

This chapter also looks at cutting-edge applications of green ML techniques to address significant problems in cellular communications. There is a severe cellular network growth network due to the advancement of green computing and the widespread usage of networks. Cellular networks may be a crucial component in assisting any organization in achieving sustainability when energy efficiency is a top concern.

This chapter is organized as follows: Section 2.2 presents the motivation study for green machine learning for cellular wireless networks. Section 2.3 discusses green ML protocols for wireless communications under the quality of service constraints. Green ML protocols in cellular systems for energy-spectral efficiency trade-off and cellular wireless base station-based energy efficiency are presented in Section 2.4 and Section 2.5. Moreover, this chapter reviews green ML for wireless access networks and power consumption minimization and reviews more efficient green ML algorithms designs in cellular radio base stations in Section 2.6 and Section 2.7. The mobile terminal energy conservation in multi-cell cellular networks and machine learning management models for green wireless networks are discussed in Section 2.8 and Section 2.9. Finally, Section 2.10 summarizes this chapter.

2.2 MOTIVATION STUDY

According to scientific data, the information and communications technology (ICT) industry's CO_2 emissions contributed 2% to the global energy consumption budget. However, as ICT applications replace jobs from the other industrial segments (such as transportation), the indirect impact of ICT on the global CO_2 footprint could be substantially more significant. Furthermore, the steep rise in the cost of energy resources in recent years has prompted big operators to try for a 20–50% decrease in energy use. The primary goal of creating mobile systems such as the Global System for Mobile Communications (GSM) and the Universal Mobile Telecommunications System (UMTS) was to improve performance.

Growing energy usage drew little attention and hence went unaddressed. Energy consumption will be a big issue for future mobile systems such as Long Term Evolution (LTE), LTE-Advanced, and 5G due to the rising energy costs and denser networks. All parts of the ICT community have recognized this obligation, including industry and academics, and new projects have sprung up worldwide, predominantly in the European Union (EU). Several EU projects aim to improve existing systems' energy efficiency and establish future design principles in all aspects of green networking [7].

2.3 GREEN ML FOR WIRELESS COMMUNICATIONS UNDER QoS CONSTRAINTS

The system design and performance of wireless communications are substantially impacted by the randomly varying channel conditions and the limited energy resources. In wireless systems, multipath fading, mobility, and a changing environment cause the received signal's power to fluctuate erratically over time and space. These erratic variations in received signal intensity have an impact on the maximum instantaneous data rate that a channel can support. Furthermore, because most mobile wireless devices have limited energy resources, energy efficiency is important. Additionally, the need for energy-efficient operation is growing, especially in times of limited energy resources, thanks to emerging wireless ad hoc and wireless sensor networks applications. Green communications, which attempt to minimize energy costs while reducing the carbon footprint of communication systems, have finally regained popularity. In conclusion, energy efficiency in wireless systems continues to be a major concern [6, 8].

Meeting certain quality-of-service (QoS) requirements is essential for delivering acceptable performance, quality, and energy efficiency in many wireless communication systems. In voice over IP (VoIP) and interactive video applications like video conferencing and streaming video in wireless networks, latency is a critical QoS parameter. As a result, in these circumstances, the information must be transferred within the allotted time frames.

On the other hand, as was previously noted, wireless channels are characterized by random fluctuations in the channel, and because of these unstable circumstances, providing QoS assurances in wireless communications is frequently challenging [9].

The key design challenge for next-generation wireless systems is how to provide the highest performance levels while preserving energy and adhering to particular QoS requirements. These systems are intended to enable dependable and robust communications anytime, wherever. Packet loss, buffer violation probability, and latency are a few QoS limitations. Information theory establishes the ultimate performance bounds and the optimal resource allocation. The availability of the channel state information (CSI) at the transmitter and receiver has been thoroughly examined from an information-theoretic approach for wireless fading channels. Studies on information theory frequently ignore latency and QoS constraints, even though they can produce impressive results [10].

The majority of communication networks' efficient implementation and operation, on the other hand, depends on meeting QoS standards. As a result, for many years, one of the primary considerations in the networking literature has been the instrumentation handling and satisfying QoS limitations. The theory of appropriate bandwidth of a time-varying source was developed to address this problem and identify the minimal transmission rate necessary to satisfy statistical QoS requirements. Queuing theories and large deviations are useful tools in this development. In this section, most of the recent green ML algorithms were briefly discussed with their impact on communication QoS performance [10, 11].

2.3.1 Effective capacity and energy consumption

The effective capacity and the bit energy measure are described in this part to introduce the basic ideas.

A. **Capacity Efficiency**

Based on the channel parameters, the instantaneous channel capacity in wireless communications varies arbitrarily. Transmission speeds for reliable communication can also vary depending on the source. By viewing the channel service process as a time-varying source with a negative rate, the source-multiplexing rule [12] can be utilized to incorporate time-varying channel capacity into the concept of appropriate bandwidth.

Effective capacity was described by the author in [13] as the equivalent of effective bandwidth. A statistical QoS demand is guaranteed by the largest constant arrival rate that a particular service process can handle, as represented by the QoS exponent, also known as effective capacity. is the decay rate of the queue length tail distribution if Q is the stationary queue length.

$$\frac{logPr(Q \geq q)}{q} = -\theta \tag{2.1}$$

As a result, $Pr(Q$ qmax) eqmax can be used to approximate the likelihood of a buffer violation for huge qmax. Greater denotes higher QoS restrictions, while smaller denotes lower QoS guarantees.

In a similar vein, $Pr(D$ dmax) edmax for big dmax, when the arrival and servicing processes decide [14], if D denotes the buffer's steady-state latency. Let S[t] be the accumulation of time and S[i], i=1, 2,... define the stochastic service process that is stationary and ergodic in discrete time. Assume that S[t] has a Gärtner-Ellis limit, which is expressed as.

$$Ac(\theta) = \frac{1}{t} E\left\{e^{\theta s[t]}\right\} \tag{2.2}$$

Here, C_E calculates the real capacity using the formula $(\text{SNR},\theta) = -\frac{Ac(-\theta)}{\theta} = -\frac{1}{\theta t} E\left\{e^{-\theta s[t]}\right\}$. The following table provides the effective capacity that streamlines the fading process and changes independently from frame to frame over the course of the frame period T.

$$C_E(SNR,\theta) = \frac{1}{\theta T}\left\{e^{-\theta TR[i]}\right\} bits/s. \tag{2.3}$$

Where, for some transmission strategies, $R[i]$ is the instantaneous service rate i^{th} frame duration's instantaneous service rate $[iT\text{: }(i+1)T]$, for the ith frame duration. This is done by making use of the fact that instantaneous rates $\{R[i]\}$ vary independently.

B. Energy Consumption

In cellular communications, mobile phones and tablets heralded the dawn of a new era of mobile computing. In order to satisfy the demands of these markets, the memory industry created new mobile-specific memories in response to the requirement for longer battery life. Increasing capacity and performance are being driven by cloud computing nowadays to accommodate heavier loads from linked devices [15].

The need for better memory performance, capacity, and power efficiency will be driven by AI/ML applications in the future, creating a number of design issues for memory systems. According to Open AI, the AI/ML training capability improved by a factor of 300,000 between 2012 and 2019, doubling every 3.43 months [16, 17].

The models and training sets for AI/ML are also becoming more prevalent, with the largest models about to surpass 170 billion parameters. The cloud-computing infrastructure based on cellular connection allows neural network model training to be divided among numerous servers, running in parallel, reducing training time, and handling colossal training sets. Furthermore, because of the value provided by these training workloads, a firm's "time-to-market" incentive encourages the completion of training runs. More memory bandwidth may be required if data transfer to the AI model becomes sluggish. There is a larger need for power-efficient and compact solutions because this training takes place in data centers with increasing space and power restrictions, which can lower costs and make adoption easier [18].

Once an AI/ML model has been trained, it can be used in the data center, at the network edge, and increasingly in IoT devices. To produce answers in real-time, AI/ML inference necessitates memory with both high bandwidth and low latency. Because inference is becoming more prevalent across various devices, affordability becomes a higher issue than data center hardware. Depending on the needs, AI/ML solutions often select between three types of memories: on-chip memory, High Bandwidth Memory (HBM), and GDDR Memory [19–21]. The two most effective exterior recollections in order to keep up with the most demanding applications, HBM and GDDR have developed through several generations. The quickest and most energy-efficient memory currently accessible is on-chip memory, but its capacity is quite constrained. For systems with smaller neural networks that only do an inference, on-chip memory is typically an acceptable option [19].

HBM is ideally suited for training large neural network models and training sets, whereas GDDR6 is well suited for both training and inference. The BSs (LTE eNodeBs) can be active processing UEs' data or idle in the cellular network domain (transmitting only downlink control signaling). In both states, the BS consumes nearly the same amount of energy. As a result, switching the BS (or any of the BS's components) off during idle periods could result in significant energy savings. However, putting this seemingly simple principle into practice is problematic because idle periods are not deterministic, and are determined by the sporadic distribution of traffic in terms of both space and time. The classification of cell approach can be used to extend the applicability of machine learning in mobile network management. A supervised learning method called categorization assigns an entity to a particular class from a list of categories based on how its features have changed over time for example relevant metrics.

The classifier builds a model that recognizes classes from fresh incoming data using a training data set for example a collection of entities and their features and their appropriate categorization collected from past data. The BSs in the RAN are the entities in this

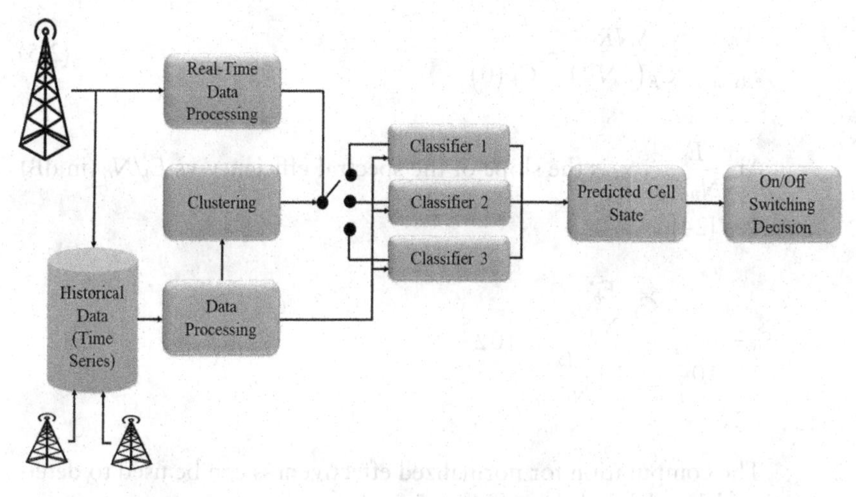

Figure 2.1 **ML-based method for mobile networks to save energy.**

scenario. Cell load parameters like the volume of traffic being transmitted and the number of assigned physical radio blocks (PRB) to UEs are among the features that are being assessed. For each fresh sample it receives from a particular BS, the classifier determines its class for the upcoming period.

The categorization produces a model that can forecast the status of a BS based on its current load with reasonable accuracy. To lessen the complexity of the system, a classification model for each BS in the RAN was developed as opposed to establishing one for each BS individually, which is impracticable when mobile network operators manage thousands of BSs. According to how similar their attributes are, things are divided into groups using the unsupervised learning process known as clustering. Figure 2.1's clustering block takes a set of features into account. The typical user count, throughput, PRB usage, and corresponding variations are calculated from historical data [22, 23].

C. **Spectral Efficiency**

The following formula can be used to calculate spectral efficiency in bits per second per Hertz if the effective capacity is divided by the bandwidth.

$$C_E(SNR,\theta) = \frac{C_E(SNR,\theta)}{B} \frac{1}{\theta TB} \left\{ e^{-\theta TR[i]} \right\}, \tag{2.4}$$

The bit that uses the least amount of energy is $\frac{E_b}{N_{0\,\min}}$. Restrictions are available from [24] under QoS.

$$\frac{E_b}{N_{0\,min}} = \frac{SNR}{C_E(SNR)} = \frac{1}{C_E(0)} \tag{2.5}$$

At $\frac{E_b}{N_{0\,min}}$, S_0 is the slope of the spectral efficiency vs E_b/N_0 (in dB) curves [24].

$$S_0 = \frac{C_E \frac{E_b}{N_0}}{10\frac{E_b}{N_0} - 10\frac{E_b}{N_{0\,min}}} 102 \tag{2.6}$$

The computation for normalized effectiveness can be used to determine the wideband slope [24].

$$S_0 = \frac{2\left(C'_E(0)\right)^2}{C''_E(0)} log\, log 2 \tag{2.7}$$

The first and second derivatives of the function C_E (SNR) in bits/s/Hz at zero SNR are denoted as $C'_E(0)$ and $C''_E(0)$, respectively.

From the first to the fifth generation (5G) of cellular technologies, they have gradually advanced [25]. Around 8.4 billion linked devices and 2.7 billion smartphone users were present in 2017. According to predictions, there will be 20.4 billion linked devices and 3.5 billion smartphone users by 2020 [26]. With more people using smartphones, wearables, and IoT devices, it is getting harder to provide high data speeds, coverage, and low latency. Furthermore, the addition of hardware in accommodating new applications and requirements increases energy consumption in successive generations. It is projected that 5G will significantly increase this traditional energy usage trajectory. The networks are becoming more energy-hungry because of the necessity to support the higher data rates and a larger number of devices. Studies suggest that 5G consumes four times the amount of energy as 4G [1]. Mobile networks currently utilize 0.5% of all energy on the planet [27].

Recently, there are many studies related to the energy efficiency for green radio access to improve the power supply in cellular networks. Different technologies help to support potential technical demands for environmentally friendly, energy-saving, and sustainable wireless access networks. By expanding the signal bandwidth, the multiplexing factor, the frequency reuse factor, or by reducing the transmission power, these technologies make cellular networks more energy-efficient, enabling the formation of numerous distinct cellular network architectures (see Figure 2.2). The 5G network

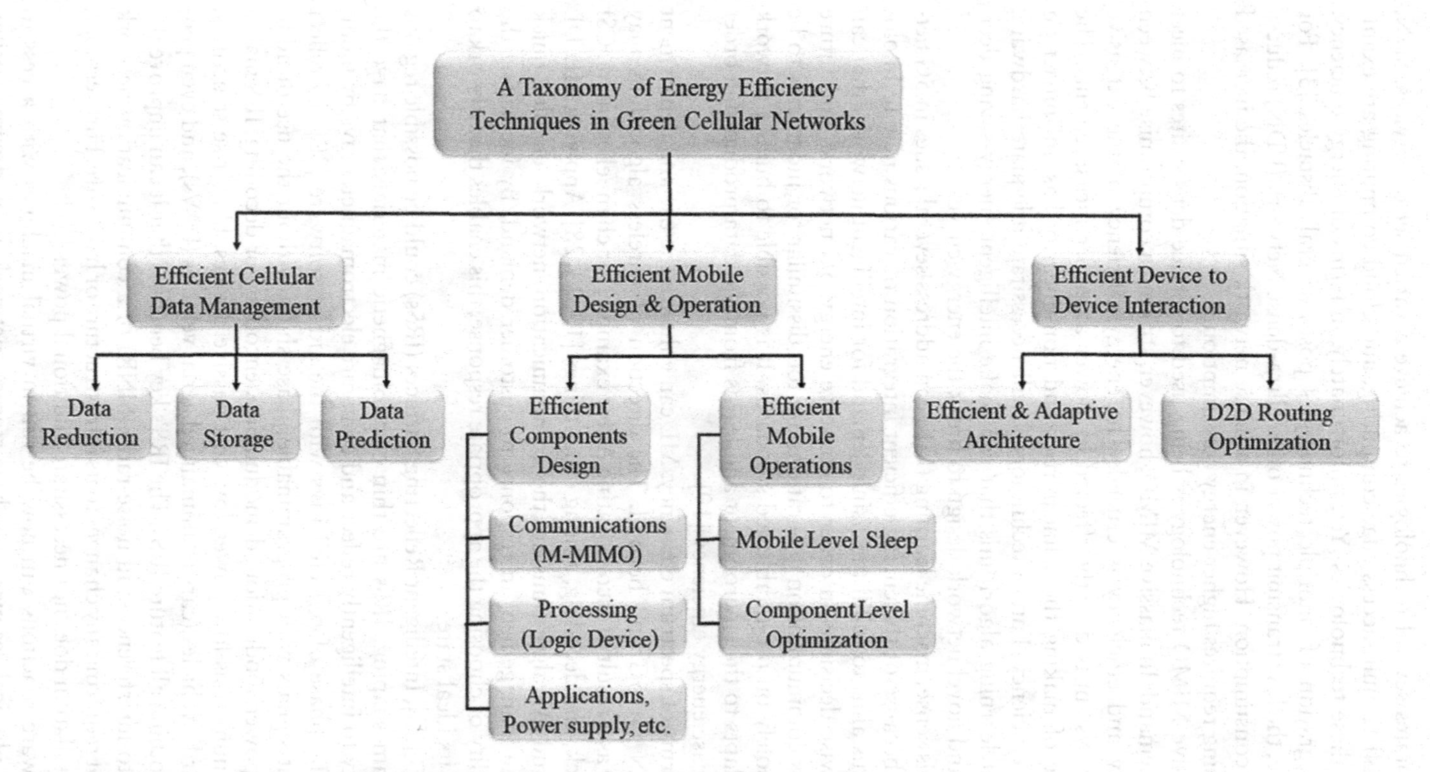

Figure 2.2 Energy-efficiency technologies for green radio access networks.

incorporates several technologies to achieve a varied set of services. SDN, UDN, NFV, multi-access edge computing, and cloud computing are examples of these technologies. Yet, unfortunately, in terms of energy efficiency, the integration of multiple technologies poses several obstacles [3]. For instance, the low transmit power of an Ultra-dense Network (UDN) reduces energy consumption. However, in a dense network situation, the increased processing requires higher energy consumption.

Massive MIMO technology is being used to service denser sites to meet rising demand. In massive MIMO, however, striking a compromise between linearity and efficiency is crucial. The energy efficiency of the extensive MIMO system is directly affected by power amplifier measurements. The expense of making them linear rises and non-linearity has an impact on energy efficiency. Future technologies will necessitate adequate hardware, efficient learning algorithms that can make intelligent energy-saving decisions, and novel network design to break the energy curve.

In this sense, machine learning (ML) can address several issues in 5G networks because of the energy-efficient integration of various new technologies. This aims to fulfill the rising demand for intelligent networks that can make wise decisions in order to conserve energy. 5G networks and future wireless communication generations are too dissimilar for judgments to be made solely on pre-established standards. It is possible to build a network that adapts to the environment and learns from the data produced in order to increase energy efficiency [19].

In terms of spectrum efficiency, ML can enhance the use and deployment of RAN for 5G and beyond. The direction of wireless algorithms may change as a result of incorporating AI for example, for channel estimate, CSI feedback, and decoding, among other things [19, 23, 28]. Applying ML, DL [29], and AI algorithms to the communication network enables quick resource management in response to customer demand. By increasing the possibility of choosing the appropriate response, this enables the network to stay in its ideal state.

In 6G [30], Intelligent Reflecting Surfaces (IRSs) could be possible regions for beam shaping. IRSs are thin electromagnetic materials that have the capacity to intelligently reflect and configure electromagnetic rays by modifying the phase of reflected rays with the aid of software [30]. To reflect incident signals with programmable phase shifts without the use of additional power, modulation, demodulation, encoding, or decoding, IRS makes use of numerous low-power, low-cost passive devices. High-rise structures, billboards, vehicles (cars or unmanned aerial vehicles (UAVs), and even people's clothing all feature IRSs. The IRS's key benefit is that it can improve the signal-to-interference-and-noise-ratio (SINR) of a communication network without requiring any changes to its infrastructure or hardware. Furthermore, the installation does not necessitate additional power [31].

Software functions can now be run in virtual machines with access to shared physical resources, such as storage, networking, and computation, thanks to network function virtualization. Containers are used to instantiate a variety of functions within the same virtual machine. To handle shifting network demands, such as offered services and network traffic, the virtual computers can be dynamically instantiated. Just a few services that can be virtualized include load management, mobility management, baseband processing unit services, and evolved packet core (EPC) activities. Unlike traditional mobile networks, network function virtualization (NFV) allows virtual network functions (VNFs) to route packets of each service between them [32]. Furthermore, the routing services are given at a low cost. Additionally, even when a new VNF is added or removed in response to traffic demands, routing and traffic flow are kept uninterrupted.

2.3.2 Low-power regime energy efficiency

As described in the preceding section, the minimum bit energy is obtained as $SNR = \frac{P}{N_0 B} \to 0$. As a result, energy efficiency increases whether one operates in the low-power regime, where P is small, or the high-bandwidth regime, where bandwidth B is great. These two regimes achieve comparable results in terms of Shannon capacity. Hence they can be considered equivalent.

Researchers started concentrating on energy-efficient communication about ten years ago. Energy efficiency has become even more crucial due to the quick development of communication technology and the rise in demand for high-data-rate applications [32]. EARTH [12, 33], eWIN [34, 35], OPERA-Net [36, 37], and Green Radio [38] are a few research projects that have been started in the recent decade to improve the energy efficiency of wireless communication networks. EARTH was a major success. It was committed to developing novel deployment strategies, network management software, and energy-efficient hardware. According to studies, the employment of different radio access technologies and energy conservation by reducing cell size and adding sleep modes can help overcome load variations. Base station sleep duration, cell size, adaptive load variation algorithms, relay and cooperative cellular communication, resource allocation strategies, and other radio access techniques have all contributed to making wireless communication more energy efficient at times.

In the future generation wireless network, a technique to turn off dormant cells in low traffic environments using an analytical model can move the trend toward green communications (see Figure 2.3). Pico cells [39] and femtocells [40, 41] reduce route loss and increase network energy efficiency. Green cellular architecture, on the other hand, allows for less radiation from user terminals. In the last decade, relays and cooperative communications have received much attention for their ability to reduce power usage. In the recent past, BS energy conservation has also gotten much attention.

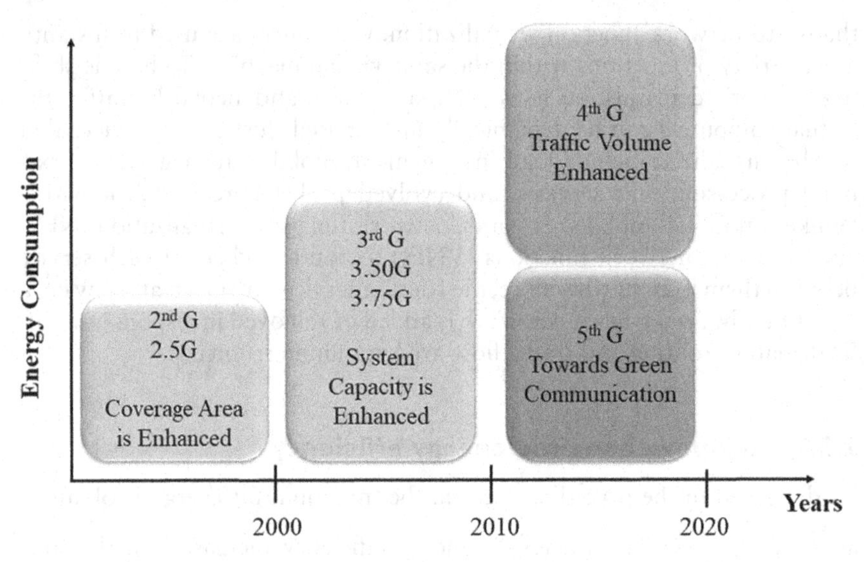

Figure 2.3 The shift toward green communications.

Variations in traffic volumes at base stations waste much electricity. China Mobile's use of dynamic power saving minimizes power consumption by 27% [42].

2.3.3 The wideband systems energy efficiency

The performance at high bandwidths is investigated in this section, while the average power $\underline{P}$ is held in place. The impact of θ is investigated on $\frac{E_b}{N_{0\,\min}}$ and the slope of the wideband S_0 in wideband. It is essential to keep in mind that as the bandwidth increases, the average signal-to-noise ratio $\text{SNR} = \underline{P}/(N0B)$ and a reduction in spectral efficiency. In addition, the coherence bandwidth increases with increasing bandwidth due to multipath sparsity if the wideband channel is fragmented into sub-channels, each having a bandwidth equal to the coherence bandwidth. In contrast, the analysis is valid when the number of sub-channels remains constrained. The lowest bit energy and wideband slope values can be determined from the Section 3.2 results when the coherence bandwidth and the number of sub-channels climb inexorably with increasing bandwidth by letting β and hence $\beta = \theta T$ tends to infinity when $\theta > 0$. Sections 3.4 and 3.5 go over the influence of multipath sparsity and the breakdown of the wideband channel into narrowband sub-channels.

Machine learning approaches are now available for a network with the arrival of SDN. Machine learning in network management is made possible by historical data and online learning ideas. The ongoing and continuous nature of the network management problem lends itself well to reinforcement learning. In 5G heterogeneous cloud radio access networks, RL is also employed for energy-efficient resource allocation. The Q-learning approach can be used to allocate wideband resources [12, 13].

Because of the 3GPP NR networks' increasingly complicated service needs, some of which have been covered in earlier sections, and their varied and strict service requirements, traditional approaches to network operation and optimization are no longer appropriate. Such methods for characterizing natural habitats rely heavily on theoretical presumptions and expert knowledge. They cannot handle the intricacies of real-world scenarios, which involve several factors, defects, and stochastic and non-linear processes, and as a result they do not scale well. To close this gap and provide 3GPP NR networks the intelligence needed to choose the best operation areas, equipment manufacturers and MNOs have started to incorporate ML-based features into their devices [43, 44].

Today, supervised and unsupervised learning approaches [45], two distinct branches of machine learning, are frequently used to first model the behavior of 5G networks before making informed decisions and/or making predictions on complex scenarios [46]. This is particularly accurate in terms of energy efficiency. As previously noted, reducing 5G energy consumption is a significant network issue that is mostly dependent on intricate BS and UE distributions, fluctuating traffic demands, wireless channels, as well as covert network trade-offs. For the 5G network's MIMO, lean carrier, and advanced sleep modes to be customized, it is essential to comprehend and predict UE behavior, requirements, and their evolution across time and location. In addition, to ensuring that the lowest possible amount of energy is used to meet the specific communication needs of the user's equipment.

In addition, reinforcement learning (RL) [32] is being extensively researched in order to improve the performance of 5G networks in general and energy efficiency in particular due to the dynamic nature of wireless networks and the lack of network measurement data for all network procedures and all possible configurations. For instance, turning off network components is a complex combinatorial task. RL agents can help the network engage with its surroundings and discover the most effective (de)activation techniques, hence lowering the overall energy consumption of the network. However, such learning might be offset by massive computing and storage capacities, an unduly long exploration period, and subpar 5G network performance [47]. Furthermore, it can be difficult to modify an ideal policy developed from a small number of distinct system configurations when network parameters are continually changing.

2.3.4 Fixed-rate and power transmissions

The transmitter provides data at a constant rate and power and is thought to have no channel knowledge. In this fixed-rate scenario, a two-state

(ON-OFF) transmission paradigm is adopted, in which information is consistently sent at a fixed rate in the ON state and is not transmitted at all in the OFF state. We investigate the wideband regime for both sparse multipath fading. As bandwidth increases and rich multipath fading occurs, the number of sub-channels remains constrained, but the number of non-interacting sub-channels grows without limit [48, 49].

The wideband slope expressions and minimum bit energy are found for the wideband domain with multipath sparsity. The low-power domain, which is analogous to the wideband regime with rich multipath fading, quantifies the equations for bit energy required at zero spectral efficiency and wideband slope. For a particular class of fading distributions, it has been demonstrated that the bit energy needed at zero spectral efficiencies is the minimal bit energy for reliable communications [49].

2.3.5 Phases of training and data transmission

As noted before, it is assumed that both the transmitter and receiver are starting out with no channel side information. The transmitter is oblivious of the fading coefficients during the transmission, whereas the receiver tries to learn them through training.

A. **Training Phase**

Training and data transmission are the two phases of the system's operation. Available pilot symbols are sent out during the training phase to let the receiver make an imperfect assessment of the channel conditions. MMSE estimation is used by the receiver to determine the channel coefficient $h[i]$. The transmission of a single pilot at every T seconds is the optimal option because the MMSE estimate is based purely on the training energy and not the length of the training [50] and the fading coefficients are anticipated to remain constant across the frame duration of T seconds. It is important to remember that there are TB symbols with a duration of T seconds in every frame and that the total amount of energy available is PT. Assume that each frame contains $TB-1$ data symbols in addition to a pilot symbol. The following provides the energies for the pilot and data symbols.

$$\varepsilon_p = \rho PT, \text{ and } \varepsilon_s = \frac{(1-\rho)PT}{TB-1} \tag{2.8}$$

Where ρ is the portion of total energy used for exercise. It is important to remember that the data symbol energy is determined by evenly dividing the remaining energy across the data symbols. In the training phase, the transmitter transmits the pilot sign $x_p = \sqrt{\varepsilon_p} = \sqrt{\rho PT}$ and obtained by the receiver

$$y[1] = h\sqrt{\varepsilon_p} + n[1] \tag{2.9}$$

In this phase, the receiver uses the received signal to calculate the MMSE estimate $h_{est} = E\{h|y[1]\}$. As seen, a complex, circularly symmetric, Gaussian random variable having a mean of zero and variance $\frac{\gamma^2 \varepsilon_p}{\gamma \varepsilon_p + N_0}$, i.e., $h_{est} \sim CN\left(0, \frac{\gamma^2 \varepsilon_p}{\gamma \varepsilon_p + N_0}\right)$ [51]. The fading coefficient h of a channel can now be stated as $h = h_{est} + h_{err}$, where h_{err} is the estimated error and $h_{err} \sim CN\left(0, \frac{\gamma N_0}{\gamma \varepsilon_p + N_0}\right)$.

B. **Lower Bound for Phase and Capacity of Data Transmission**

After the training phase, the data transmission phase begins. The channel input-output connection in a single frame during the data transmission phase can be described as follows since the receiver acquires the channel estimate.

$$y[i] = hestx[i] + herrx[i] + n[i] i = 2,3,\ldots,TB \tag{2.10}$$

Determining the channel capacity in (10) is difficult [51]. A capacity lower bound is commonly derived by treating $herrx[i] + n[i]$ as Gaussian distributed noise with variance $\mathrm{E}\left\{\left|\mathrm{herrx}[\mathrm{i}] + \mathrm{n}[\mathrm{i}]\right|^2\right\} = \sigma^2_{h_{err}\varepsilon_s + N_0}$, where $\sigma^2_{h_{err}} = E\left\{\left|h_{err}\right|^2 = \frac{\gamma N_0}{\gamma \varepsilon_p + N_0}\right\}$ is the estimate error's variance. [50, 51] provide a lower limit on the capacity at any one time under these assumptions by the following.

$$\begin{aligned} C_l &= \frac{TB-1}{T}\left(1 + \frac{\varepsilon_0}{\sigma^2_{h_{err}} \varepsilon_s + N_0}\left|h_{est}\right|^2\right) \\ &= \frac{TB-1}{T}\left(1 + SNR_{eff}\left|\omega\right|^2\right) \text{Bits/s} \end{aligned} \tag{2.11}$$

where the effective SNR is

$$SNR_{eff} = \frac{\varepsilon_s \sigma^2_{h_{est}}}{\sigma^2_{h_{err}} \varepsilon_s + N_0} \tag{2.12}$$

The $\varepsilon_s \sigma^2_{h_{est}} = E\left|h_{est}\right|^2 = \frac{\gamma^2 \varepsilon_p}{\gamma \varepsilon_p + N_0}$ is the h_{est} estimate's variance. It is worth noting that the statement is derived by defining $h_{es}t = \sigma h_{est} w$, where w is a conventional complex Gaussian random variable with unit variance and zero mean, i.e., $w \sim CN(0, 1)$. Now, the study will use CL to analyze the impact of the inaccurate channel estimate.

2.3.6 Summarized insights

The energy efficiency of single-input single-output fading channels with QoS restrictions is explored in this part by analyzing bit energy levels and effective capacity as a measure of maximum throughput under specific statistical queuing constraints. Our research defined the trade-off between energy, bandwidth, and delay. We examined the trade-off between spectral efficiency and bit energy under QoS restrictions in the low-power and wideband regimes. The following situations were used to generate the analysis:

A perfect CSI is only available at the receiver, a perfect CSI is only available at the transmitter, and an imperfect CSI is only available at the receiver. Expressions for the wideband slope and minimum bit energy are discovered. This analysis was used by the perfect CSI to estimate the increased energy usage in the presence of delay-QoS constraints. Researchers may now understand the principles needed to examine the effects of different transmission systems, such as variable-rate/variable-power, variable-rate/fixed-power, and fixed-rate/fixed-power transmissions, on energy efficiency.

2.4 GREEN ML IN CELLULAR SYSTEMS FOR ENERGY SPECTRAL EFFICIENCY TRADE-OFF

In the current environment of rising energy demand and rising energy prices, energy efficiency (EE) is becoming a significant study priority in communication [52, 53]. This field has been widely researched in the past, but primarily through the lens of power-limited applications like battery-powered devices [53], mobile terminals, underwater acoustic telemetry [14], and wireless ad hoc and sensor networks are examples of these technologies. This field has been examined recently for uses requiring unlimited power, including cellular networks [24]. The shift in research objectives from power-limited to power-unlimited applications is principally driven by two economic and environmental concerns: lowering the carbon footprint of communication systems and reducing the ever-rising operational costs of network operators.

Despite this, spectral efficiency (SE) continues to be a crucial factor in assessing the performance of communication systems. The SE is a statistic that demonstrates the efficiency with which a finite frequency spectrum is used, but it does not demonstrate the efficiency with which energy is consumed. The latter will become just as important in the context of energy conservation as the former. It must therefore be taken into account in the system of performance evaluation. Maximizing the EE and the SE are competing objectives, suggesting a trade-off, or more precisely, minimizing the amount of energy used while maximizing the SE.

2.4.1 EE-SE trade-off concept

The EE-SE trade-off was initially developed for power-limited systems, and the low-power (LP)/low-SE regime accurately defined it [54]. Given the recent emergence of the EE as a key system design criterion alongside the existing SE criterion, the EE-SE trade-off will soon become the measure of choice for effectively constructing future communication systems. As was previously mentioned, EE research is shifting from applications with restricted power to those with unlimited power. So, as pioneered by the studies in [1, 26], the idea of the EE-SE trade-off needs to be generalized for power-unlimited systems and precisely stated for a wider range of SE regimes.

The EE-SE trade-off was initially developed for systems with power constraints, and it was precisely defined for the low-power (LP)/low-SE regime [54]. Given the recent introduction of the EE as a key system design criterion alongside the existing SE criterion, the EE-SE trade-off will soon become the measure of choice for efficiently constructing future communication systems. EE research is shifting from power-limited to power-unlimited applications, as was previously stated. The idea of the EE-SE trade-off needs to be expanded upon for power-unlimited systems and precisely stated for a wider range of SE regimes, as pioneered by the works in [1, 26].

In order to fully understand the total amount of power used in a cellular system and to accurately assess its EE, a realistic power consumption model (PCM) for each node must be developed. Examples of such PCMs include those suggested in [34, 36] for different types of BSs as well as for backhaul lines [35, 37]. The EE-SE trade-off equation may be reduced and expanded to include these actual PCMs, giving it a reliable and uncomplicated metric for measuring performance.

Due to the dynamic nature of network traffic because of user mobility, multi-tier BSs can be programmed to turn on and off to save energy [36]. If the BSs work status changes, the user association information should be updated to ensure a reliable connection. Because of this, it is important to carefully design the work state of BSs in order to conserve energy while yet ensuring that QoS requirements are met. Scheduling BSs states can perform better when heterogeneous approaches and machine learning are combined, as shown in Figure 2.4. The widely utilized DL approaches will probably be incorporated into the BS deployment design because the DL has demonstrated better accuracy rates and more sophisticated policy seeking capabilities in numerous researches.

Making the EE-SE trade-off into a CFE or CFA, as shown at the link level, can compromise the EE-SE trade-off in a cell. Understanding how to reduce consumed power while maintaining an acceptable SE or quality of service is

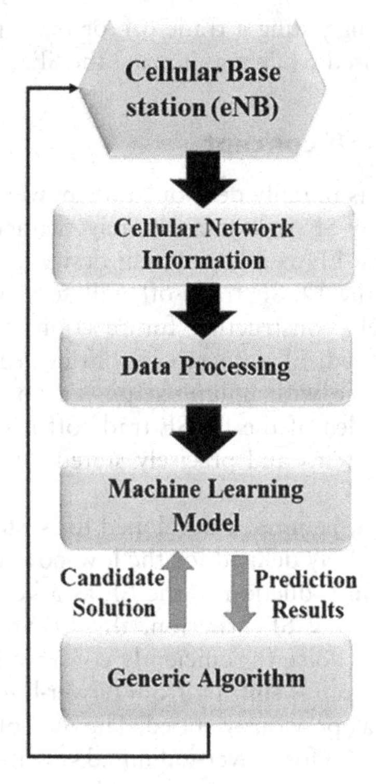

Figure 2.4 ML for intelligent BS deployment.

the first step in this process (QoS). Additionally, the idea of a trade-off is inextricably tied to optimization, as seen in Figure 2.5. At the cell level, where various users compete for a finite number of resources, optimization-based techniques, such as link adaptation [56] and resource allocation [44] algorithms, have been studied extensively. Generally speaking, an optimization method works best for a certain problem with a particular set of criteria; altering the criteria will probably change the problem and the strategy. Up until recently, the most often used parameter for creating useful resource allocation algorithms was the rate of SE.

To enhance resource allocation and enable QoS, fairness has been employed as a criterion. It does, however, frequently work in tandem with SE. Instead of being a criterion, power has been used in resource allocation as a restriction. Resource allocation based on EE is becoming more and more prevalent as EE becomes a core system design criterion, especially in the uplink of a single cell system to increase UT battery autonomy [58–60]. Additionally, resource allocation is not exempt from the present research trend that focuses on battery-limited to inexhaustible power applications. A strategy for maximizing the EE-SE trade-off in the scalar broadcast channel, for instance, was presented by the authors in [61]. In contrast, the authors

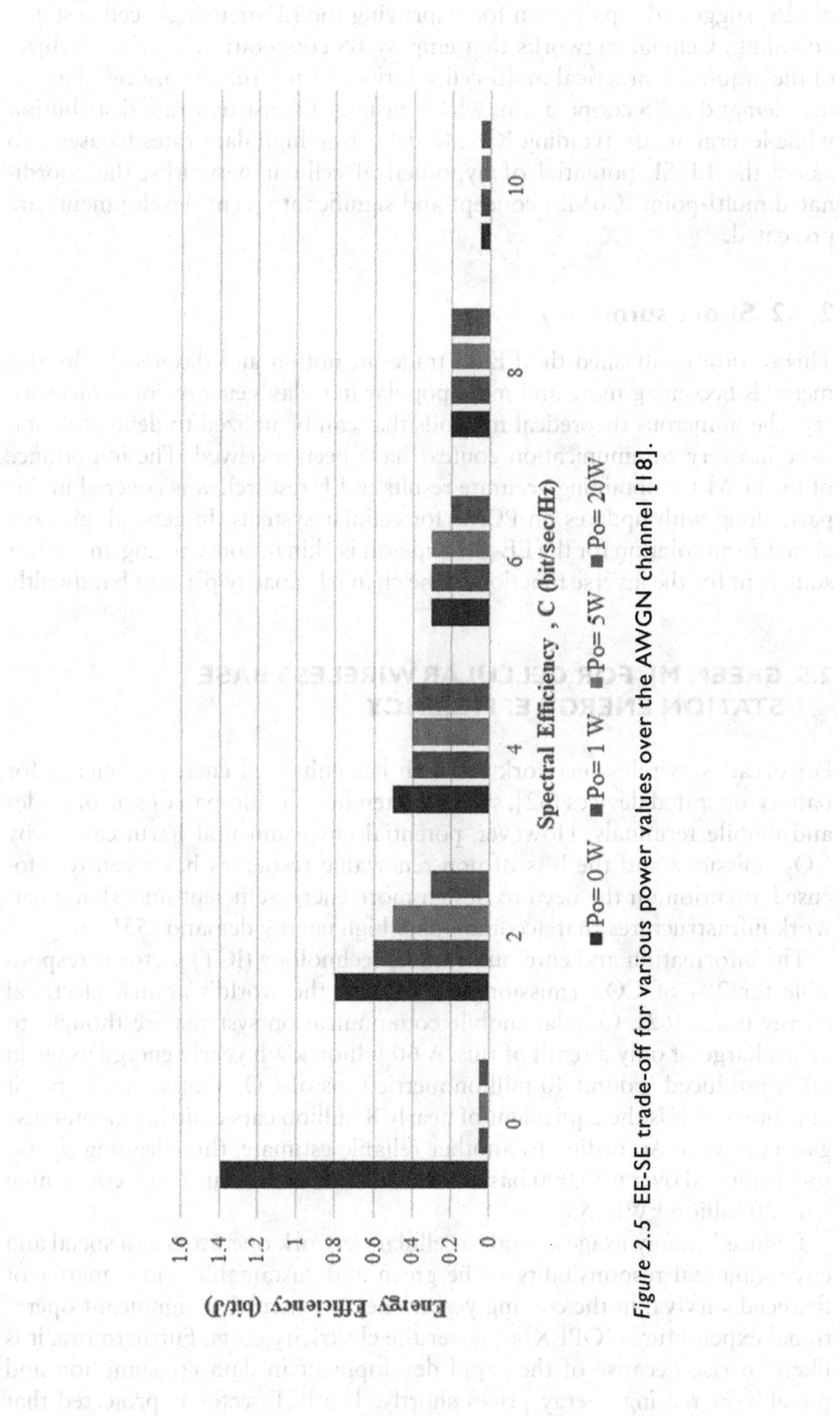

Figure 2.5 EE-SE trade-off for various power values over the AWGN channel [18].

of [28] suggested a paradigm for improving the EE in a single-cell system's downlink. Cellular networks that employ BS cooperation were the subject of the inquiry. A practical multi-cell solution for meeting future cellular system demand is BS cooperation, which ensures a consistent rate distribution while leveraging or avoiding ICI and delivering high data rates to users. To assess the EE-SE potential of hypothetical cellular networks, the coordinated multi-point (CoMP) concept and significant recent developments are presented.

2.4.2 Short summary

This section established the EE-SE trade-off notion and discussed why this metric is becoming more and more popular in today's energy-conscious society. The numerous theoretical methods that can be utilized to define this statistic in every communication context have been reviewed. The importance of the PCM for obtaining accurate results in EE research was covered in this part, along with updates on PCMs for cellular systems. In general, giving a closed-form solution for the EE- SE trade-off is akin to constructing an explicit statement for the inverse function of the channel capacity per unit bandwidth.

2.5 GREEN ML FOR CELLULAR WIRELESS BASE STATION ENERGY EFFICIENCY

For decades, wireless network research has enhanced energy efficiency for battery-operated devices [52], such as extending the lifespan of sensor nodes and mobile terminals. However, potential environmental harm caused by CO_2 emissions and the loss of non-renewable resources has recently refocused attention on the need to design more energy-efficient underlying network infrastructures that accommodate high energy demand [53].

The information and communications technology (ICT) sector is responsible for 2% of CO2 emissions and 3% of the world's annual electrical energy usage [62]. Cellular mobile communication systems are thought to be in charge of only a tenth of this. A 60 billion kWh yearly energy usage in 2008 produced around 40 million metric tons of CO2 emissions. To put it in context, this is the equivalent of nearly 8 million cars emitting greenhouse gases per year. According to another reliable estimate, three leading operators deployed over 600,000 base stations (BSs) in China in 2007, consuming over 20 billion kWh [53].

Reduced energy usage is vital to cellular network operators as a social and environmental responsibility to be green and sustainable and a matter of financial survival in the coming years. They are incurring significant operational expenditures (OPEX) to cover the electricity costs. Furthermore, it is likely to rise because of the rapid development in data consumption and possibly increasing energy prices shortly. The ICT sector is projected that the energy consumption grows at 15 to 20% per year, doubling every five years. According to ABI Research [3], this will result in a total cellular network OPEX of $22 billion in 2013. As a result, taming the spiraling OPEX is critical to operators' long-term success.

2.5.1 Green cellular network design overview

Due to these requirements for energy reduction, operators have been exploring for ways to improve the energy efficiency of various cellular network components, including mobile terminals [14], BSs, and mobile backhaul networks. According to reports, BSs account for 60–80% of total power consumption in cellular networks, making them the main energy consumer [24]. As a result, the focus of this chapter is on energy-efficient cellular BS operation, which is referred to as running a green cellular base station. Green cellular networks can achieve their goal in several ways as follows.

A. **At the component level**
   Novel designs and hardware implementations include energy-efficient power amplifiers [25] and fan fewer coolers, or even natural resource-based cooling [26].
B. **At the link level**
   Energy-aware transmission and resource management techniques, such as power control [1] and user association [27], are implemented at the link level.
C. **Network-level**
   Topological techniques from deployment through operation include intelligent deployment at the network planning stage using micro BSs or relays [33–36] or traffic-aware dynamic BS on/off [24, 50, 53, 63, 64].

The investigation of the whole problem space is done based on the following two axes, as shown in Figure 2.5: what can be managed over the time scale, and what performance needs to be compromised to save energy. Since the core of green BS design is to fine-tune the trade-off between energy and efficiency, they are evaluated by many metrics, and the phrase "sacrifice" has been adopted.

In Figure 2.6, the problem space of green cellular networks is illustrated depending on the following metrics.

A. **Timescales and control**
   The control systems in question, in addition to their operating time schedules, characterize how to save energy in cellular networks. For example, the deployments of energy-aware base stations can take months or even years, whereas transmitting power control of mobile terminals or BSs can take milliseconds.
B. **Greening cost and efficiency metric**
   Throughput, latency, and blocking or outage likelihood are major system performance parameters of cellular network applications that

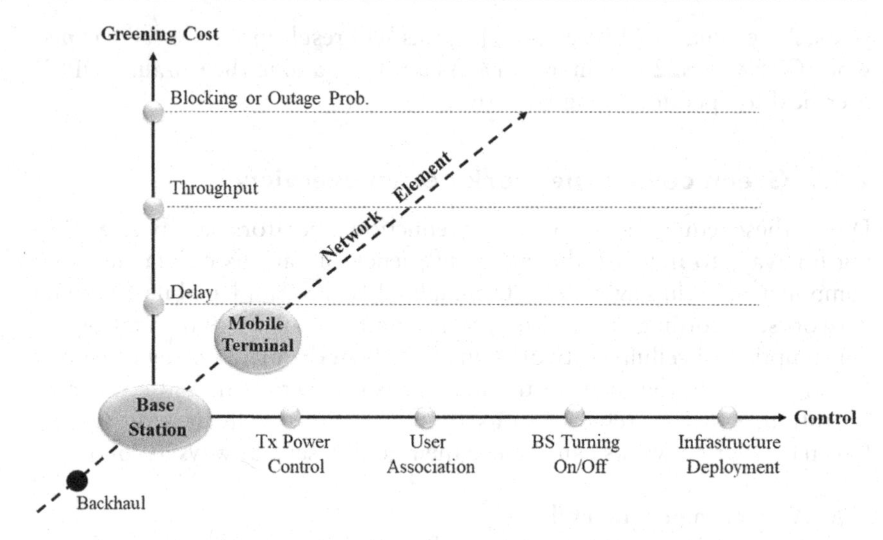

Figure 2.6 Problem space of green cellular networks.

are also compromised by power reduction. For best-effort data traffic, throughput is a good statistic to use, whereas latency and blocking/outage probability are more appropriate for real-time traffic [66].

2.5.2 Heterogeneous deployment with energy consciousness

Determining the site locations, the number of, and the specific macro/micro/pico/femto type BSs that must be installed in an energy-efficient manner is the main challenge of energy-aware deployment. The sets of macro and micro base stations, represented by BM and Bm, respectively, are taken into consideration in the model of a deployment plan for micro base stations. The section uses the subscript M for macro BSs and the letter m for micro BSs. The index of BSs is denoted as b∈B = BM ∪ Bm. The primary focus is on downlink communication, a primary method of mobile Internet usage, i.e., from BSs to mobile terminals (MTs). However, several features of the research can also be applied to the uplink [67].

For the model with links, Eb(x) = pbgb(x), where *pb* signifies the transmit power of BS *b* and *gb*(*x*) denotes the channel gain from BS *b* to location *x*, including path loss attenuation, shadowing, and other factors if applicable. On the other hand, fast fading is not considered because the time scale for measuring *gb*(*x*) is much more significant. As a result, the signal to interference plus noise ratio (SINR) at position *x* is:

$$\Gamma(x,B)=\frac{E_{b(x,B)}(x)}{\sum_{b\in B,b\neq b(x,B)}E_b(x)+\sigma^2} \tag{2.13}$$

where σ^2 are noise power and b(x, B) is the index of the BS at position x that produces the strongest signal, i.e., b(x, B) = argmax b ∈ B Eb(x). Following Shannon's formula, spectral efficiency at location x is given by C(x, B) = log 2 (1 + Γ(x, B)). The area spectral efficiency (ASE) was initially proposed in [68] and is used as a performance metric for spectral efficiency in a given area. According to this definition, the total spectral efficiency over reference region A is:

$$S(B) = \frac{\sum_{x \in X} C(x,B).\Pr(x)}{|A|} \left[\text{bps/Hz/m}^2\right] \tag{2.14}$$

where $Pr(x)$ is the likelihood that the MT will be found at position x. X is the set of locations included in the area A satisfying $Pr(x) > 0$ for all $x \in X \subset A$. For the sake of simplicity, it is assumed that the user distribution is homogeneous in this section. The discrete set X in this case is a rectangular lattice with a modest grid size, and each location has the same probability [67].

2.5.3 User association and base station on/off

Remember that we have only examined energy-aware deployment thus far; it could take months or even years to resolve an offline issue. The next challenge is figuring out how to use the BSs effectively for energy conservation during the off-peak period after they have been deployed. The answers to the operation problem should be online distributed algorithms in order to be used in practical systems. The operation problem has been the subject of various literary attempts [51, 69, 70]. A simple greedy-on and greedy-off BS switching on/off algorithm and an ideal energy-efficient user association policy are proposed as an effective but practical solution for the total cost reduction challenge that offers a flexible trade-off between flow-level performance and energy consumption.

2.5.4 A case study in quantitative analysis

The motivational review illustrates the idea of how much energy may be saved in cellular networks by turning down redundant BSs [71]. To quantitatively evaluate prospective energy savings, two sets of actual data are integrated. The data consists first an anonymous one-week temporal traffic trace from a cellular provider in a major city, as shown in Figure 2.7, and the second is BS locations, with a total of 139 BSs in 128 locations in a 3.53.5 km area.

There could be a significant redundant overlapping in cellular coverage, which assumes a perfect circular range of 700 meters for each BS. Individual operators can save between 8 and 22% of energy in an urban deployment when the temporal and spatial data sets are combined. By pooling BS

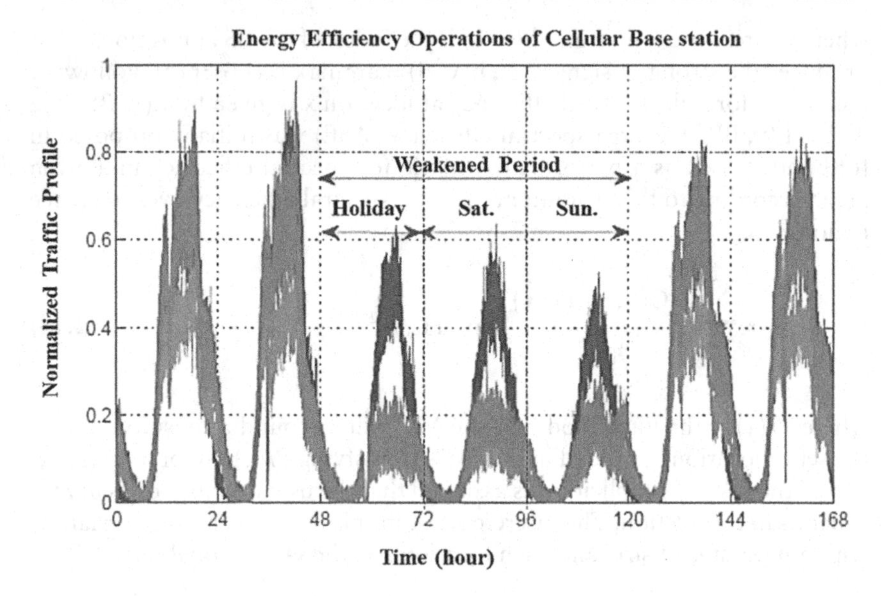

Figure 2.7 An example of temporal traffic traces from a metropolitan area cellular operator over a one-week timeframe.

resources, a total reduction in energy consumption of almost 29% can be realized. If the single BS power is between 800W and 1500W per year, this translates to a total regional electricity savings of between 78000 and 530000 kWh. The cost of electricity in USD per KWh for commercial use is approximately 0.068, according to [72]. For residential properties, is roughly 0.116, resulting in annual savings for running BSs of $5000 to more than $61000. These emissions, according to [73], range from 53.7 to 365 metric tons of CO2. This represents a significant reduction in both greenhouse gas emissions and operating costs.

2.5.5 Open issues and alternative routes to green cellular networks

Researchers need to focus more on enhancing green energy for cellular networks by emphasizing a number of open concerns and other potential solutions, despite recent articles focusing on increasing the energy efficiency of green cellular networks. The following examples [74] show a new paradigm and architecture for green cellular networks.

2.5.5.1 Cooperation between multiple operators

There is also potential for large energy savings when many operators pool their BSs and accept one another's traffic as roaming traffic while the BSs are off [60, 75]. This kind of operator cooperation to ensure pooled coverage is especially helpful in metropolitan areas when each operator deploys a dense deployment. Similar research was done by authors in [75], who looked at the straightforward configuration of a single cell with two operators and demonstrated that there can be significant energy savings in such a setup. When attempting to achieve teamwork, physical sharing may not be a crucial factor. BSs in urban areas are frequently nearby, if not actually co-located. The main barrier is the increased network operating complexity brought on by fine-granularity roaming, including cross-operator user authentication and billing. This leads to a variety of issues that can be researched from an economic and game-theoretical perspective. For the sake of environmentally friendly operations that also help society in other ways, it is important to investigate whether such operator agreements could be exploited to create oligopolies that hurt consumers and how they can be prevented.

2.5.5.2 Deactivation or slowing of components

As was indicated in the last section, shutting off those BSs that aren't in use can save a lot of energy. In order to prevent coverage gaps, a certain group of BSs shouldn't be turned off despite low traffic. A worse user experience could result from turning off BSs as well, such as increased uplink power usage for file uploading. Component-level strategies [76, 77] that are more conservative while still decreasing energy consumption have recently been developed due to the technical challenges of totally turning off BSs. In [76], the number of radio units and transmit antennas in BSs were controlled by turning them back on during times of low load. How a component-level slowing approach could be included into BS operation was examined by the authors of [77]. An energy-saving technology called dynamic voltage frequency scaling, sometimes known as "speed scaling," enables a central processing unit (CPU) to change its speed in response to incoming processing demands.

2.5.5.3 Coordinated multipoint transmission and reception (CoMP)

Although some BSs are disabled during periods of low activity, maintaining coverage and service quality is the fundamental restriction of green cellular operation. The coverage area of the remaining BSs can be expanded by raising transmit power when some BSs are disabled. Another unusual tactic that can aid in maintaining coverage is the use of coordinated multi-point transmission and reception (CoMP), which is being investigated in the context of LTE-Advanced systems [79]. In order to enhance signal quality, CoMP, a macro-diversity solution based on network MIMO, coordinates transmission and reception at nearby cells. CoMP can boost cell-edge throughput and/or high-data-rate coverage, thereby enabling systems to turn off more BSs [80]. The key hurdles to overcome are overheads such as message exchanges between collaborating BSs through backhaul and extensive signal processing at the BSs. It is critical to look into CoMP's net-energy efficiency [81], which considers additional energy usage for such overheads.

2.5.5.4 Relaying with many hops

Future cellular networks, such as LTE-Advanced systems, will face challenging constraints as demand for coverage and capacity grows. It is sometimes more cost-effective to build relays rather than to deploy more BSs to meet such requirements. Route loss between BSs and MTs can be greatly diminished by dividing a lengthy weak single-hop link into shorter, strong multihop links. They consume little power because of their compact size. Featuring wireless backhaul they allow more deployment flexibilities and are far less expensive than installing new BSs with a fixed wired backhaul.

Furthermore, multihop relaying will undoubtedly be beneficial in ensuring that the dynamic shutting down of BSs does not result in coverage gaps while saving energy. Relays are one of the future technologies for green cellular networks because of all of these factors. Relaying in cellular networks has recently been the subject of study [75] that looks at how relays might improve spatial reuse and so offer the necessary data rates while using less energy. A relay caching system [82] and the joint optimization of relay location and the sleep/active problem [37] increase energy efficiency, especially for multimedia applications.

2.5.5.5 Control of transmit power

Small cells are susceptible to interference, which has a significant negative impact on their potential gain [32]. With a combination of macro and small cells, it is anticipated that current and future cellular networks will become more heterogeneous. Interference is a key barrier that can limit the potential gain of small cells, and its pattern is quite varied. Users that experience low throughput due to severe interference are more likely to do so when the number of tiny cells and users at cell boundaries both increase. The research community concentrated on dynamic interference management (IM) algorithms that perform dynamic BS power control based on planned users over several neighboring channels after starting with static IM methods with particular reuse patterns. Here the reusing of the patterns is traditional frequency reuse, fractional frequency reuse (FFR) [47], or variations [83, 84]. From what is discussed in this section, it is clear that a number of international green network projects and consortia that have been implemented by research institutes, universities, and companies in recent years have contributed to controlling the transmission power of cellular communications.

2.5.6 Summary

Cellular base stations in mobile networks consume large energy approximated to 70% of total energy of overall network operator. The use of renewable energy sources on the network of base station sites based on 5G technology has new features to improve energy efficiency. 5G technology has new features to improve energy efficiency. The 5G technology has also accompanied the development of batteries for devices that enhance energy management. In proportion to the energy consumption that is used by communication devices, energy efficiency and consumption optimization techniques help operators reduce consumption according to their own energy consumption assessment by percentages by measuring the power consumption of their network before deciding on a location focus on saving energy. The massive MIMO technology splits the power consumption of 5G base stations into two parts antenna module and baseband module which reduce the total base station consumption as reasonable.

AI technologies provide energy-saving solutions in a multi-network collaborative system that saves basic data such as configuration and performance statistics to network files. The AI algorithm helps to manage the operation of some network functions according to the requirements of QoS while achieving the goal of reducing energy consumption. AI assists in a number of stages, including evaluation and design network where algorithms are used that analyze energy-saving scenarios and make a prior evaluation of energy-saving benefits. Then the algorithms check the basic effect of the energy-saving solution according to the performance indicators (KPIs). The results of the network analysis are used in the implementation phase of the solution on the network, taking into account the third party KPIs of the network to ensure that the service experience is not affected and improved.

2.6 GREEN ML FOR WIRELESS CELLULAR ACCESS NETWORK

The CO_2 emissions of the ICT industry contributed 2% to the entire CO_2 budget of worldwide energy consumption, according to research findings [85, 86]. However, as ICT applications replace jobs from other industrial segments such as transportation, the indirect impact of ICT on the global CO_2 footprint could be substantially greater. Furthermore, the steep rise in the cost of energy resources in recent years has prompted significant operators like Vodafone and Orange to strive for energy usage reductions of 20% to 50% [87]. Performance enhancement was the main focus of the development of mobile systems like the Global System for Mobile Communications (GSM) and the Universal Mobile Telecommunications System (UMTS). Growing energy use received little attention and was therefore not addressed.

However, because of rising energy prices and denser networks, energy consumption will be a significant issue for future mobile systems like Long Term Evolution (LTE) and LTE-Advanced. All segments of the ICT community, including business and academia, have acknowledged this duty. As a result, new initiatives are being created all over the world, mostly in the EU.

A number of EU projects seek to improve the energy effectiveness of current systems and discover potential design ideas for completely green networking in the future. The long-term objective of the TREND project, which stands for "Towards Real Energy-efficient Network Design," is to establish the foundation for a new, all-encompassing approach to energy-efficient networking by integrating the EU research community into green networking. In order to conserve energy when wired network devices are not being used (to the fullest extent possible), the Low Energy Consumption Networks (ECONET) project suggests using dynamic adaptive technologies. Several EU initiatives seek to improve the energy effectiveness of current By combining cognitive radio and cooperative strategies to maintain the required data rate and Quality of Service (QoS), the Cognitive radio and Cooperation strategies for POWER (C2POWER) Saving in Multi-Standard Wireless Devices project seeks to develop and demonstrate energy-saving technologies for multi-standard wireless mobile devices.

In the case of cell deployment strategies, small cells are frequently regarded as lowering system power consumption. Each BS, on the other hand, uses power for both fundamental functioning and signaling. The most energy-efficient Inter-Site Distance (ISD) between BSs is thus non-trivial to choose. Traffic demand, antenna height, varied power class BS, complex power consumption behavior with traffic load, and other factors all come into play [87]. The impact of several power models will be investigated initially. The thorough EARTH power model [24], generated from actual hardware components, is used to calculate the best ISD for perfect hexagon deployments. Following that, more complex deployments with various cell sizes are considered.

Cells with smaller transmission ranges save a lot of transmit power because of the significant path loss over the air. On the other side, this necessitates additional base stations. Because macro sites already consume a large amount of offset power when they are empty, each extra cell increases the number of power-consuming components in the network. Therefore, the potential for cost savings for the overall network power utilization must be evaluated. In order to map RF transmit power to the overall power drawn from the power grid, a load-dependent power model is required.

With a smaller ISD, the needed cell edge rate can be reached with less than 40 W of transmit power, reducing interference from neighboring cells. As a result, we are looking at the best BS ISD when the macro base stations' transmit power is adjusted to cover 95% of the area. In the range of inter-site distances of 500 m to 1500 m, this leads to macro cell transmit powers of approximately 0.7 W to 87.3 W [36]. Near the vertices of the hexagonal

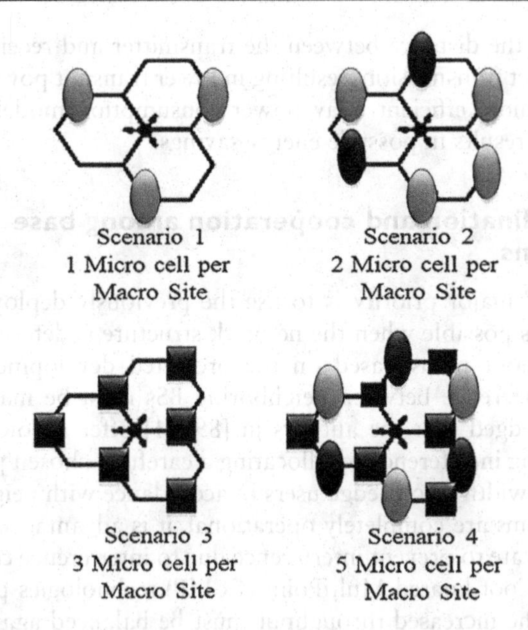

Figure 2.8 Placement of microcells in a hexagonal macro cell network.

macro cells, the signal quality from the macro base stations is relatively poor, resulting in low data rates in these locations. Remote users require more transmission resources than users near the antenna for a given data rate. Therefore, moving cell edge users into microcells intended for serving the cell edge can significantly reduce the amount of electricity used (see Figure 2.8).

The macro cells must achieve the 95% coverage objective since the microcells are regarded as an additional under-layer that provides capacity. Micro base stations have a 1W maximum broadcast power because they are designed to cover relatively short distances.

Additionally, there is a significant reduction in the amount of power used for cooling and signal processing as compared to macro cells. The LTE-Advanced study item is looking into relaying as a technology that offers more flexibility and economical deployment options by extending coverage and boosting capacity [25]. A common strategy for several radio technologies, including Wi-Fi, WI-MAX, etc., is relaying. The purpose is to divide the transmission into more manageable hops. Direct contact with user terminals may not have the same channel quality as the link between base stations (or a transmitter) and relays (or a receiver).

For two potential examples of relay deployment in a cellular system, relays can be utilized to increase capacity and/or coverage. The Spectral Efficiency (SE) and Energy Efficiency of mobile systems, for example, can both be enhanced with the use of relaying technology (EE). The SE is better because the attainable rate is comparatively higher. However, the EE is better because the distance between the transmitter and receiver is shorter than the indirect transmission, resulting in lower transmit power. This, combined with a more efficient relay power consumption model than a base station model, results in possible energy savings.

2.6.1 Coordination and cooperation among base stations

The operators' major priority is to use the previously deployed resources as efficiently as possible when the network structure is determined according to the rollout plans based on the projected development of traffic. Primarily, interference between neighboring BSs must be managed. It has been acknowledged that the authors in [89–91] offer a potential method for coordinating interference by allocating a carefully chosen portion of the available bandwidth to cell-edge users in accordance with neighboring BSs. Once the systems are completely operational, it is advantageous for BSs to actively cooperate to prevent interference due to interference constraints. At the cell edge, Coordinated MultiPoint (CoMP) technologies provide better throughput. The increased throughput must be balanced against the additional energy used by the collaboration for backhaul connections and signal processing.

FFR has been suggested as a competitive approach to reduce interference in Orthogonal Frequency Division Multiple Access (OFDMA) networks as opposed to single-frequency reuse deployments [43, 92, 93]. FFR works by dividing the entire bandwidth into two sub-bands. Users near their base station have their sub-band (inner zone of each cell). With single frequency reuse planning, this sub-band is utilized. The remaining sub-bands are occupied by users positioned near the cell edge and have a higher frequency reuse factor of (2, 3, or more) (outer zone of each cell).

The original plan was to treat all BSs the same, regardless of their traffic loads. A static solution can easily handle this situation. However, if the load between neighboring BSs is imbalanced, the overloaded cell will boost the transmit power to compensate for the lack of spectrum resources. This will result in increased interference and, in effect, a waste of energy. It is suggested that a dynamic system be utilized that tracks the number of users per cell and distributes power to each sub-band based on the relative narrowband transmit power (RNTP) indications that are present in LTE networks [70].

The primary idea behind this adaption is to provide greater bandwidth to users who are near overcrowded cells. To be effective, the adaptation must be completed as quickly as feasible. As a result, this strategy might be classified as "coordinated scheduling." Figure 2.9 depicts a simple downlink scenario with two nearby BSs, indicated as BS_1 and BS_2, respectively. Three slices of bandwidth are available: For both locations, N_0 Physical Resource Blocks (PRBs) are used with a frequency reuse factor of 1, and N_1 Physical Resource Blocks (PRBs) are used with a frequency reuse value of 0.

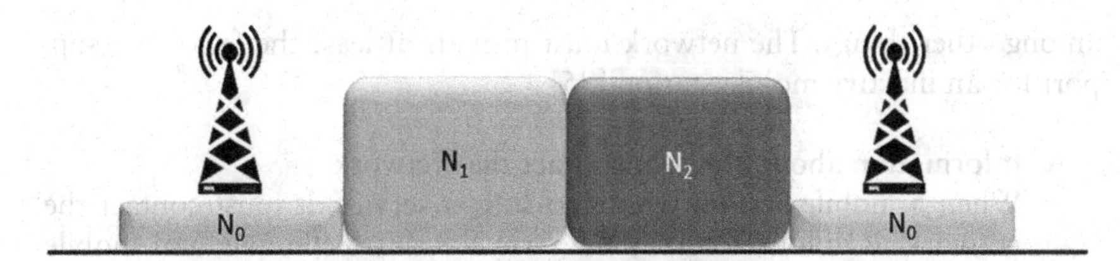

Figure 2.9 Two base stations in an adaptive FFR scenario.

Respectively N_2, PRBs are only used by BS_1 and BS_2 for users who are near the cell edge.

An approach called dynamic optimization can be used to estimate the size of each sub-band, N0, N1, and N2. Following the allocation of BS1's power and frequency resources (e.g., using the method described in [94]), BS2 assesses the interference posed by BS1 using the RNTP indicator and assigns the resources in accordance with its findings.

The total cell throughput and cell edge throughput can both be significantly increased by the CoMP approaches. However, because of the added overhead and power consumption, the strategy places a certain amount of additional burden on the system.

According to a study of present network traffic and projections for future development, BSs will continue to use just a tiny portion of their capacity [24]. On the one hand, the stringent requirements of low latency and low blocking likelihood necessitate that resources are typically underutilized even during peak hours. Furthermore, examining the daily variation of data traffic in today's networks [24, 72] reveals extended periods during the day when the network's average load is 5 or 10 times lower than peak values during busy/peak hours. Even though the actual traffic demand is significantly lower, a significant portion of the daily energy consumption is spent maintaining total system capacity. To save energy, innovative management aims to reduce such overprovisioning.

Dynamically reducing the number of active network parts to follow the daily traffic variation is a promising technique to reduce mobile network energy usage. Adaptation methods take advantage of free resources by switching the transceiver hardware to a lower-energy mode. It is critical to manage resource use in such a way that such hardware improvements are maximized.

2.6.2 Future architectures

Despite the current cellular architecture's success, numerous areas might be improved to save energy. To give existing systems more flexibility and intelligence, it is necessary to change, re-plan, or even completely redesign them. This requires figuring out how to detach data from system information, expand existing systems with multihop, and benefit from network coding, among other things. The network must provide at least the following support for an inactive mobile station [95]:

A. **Information about How to Contact the Network**
When a mobile station wishes to start a service, it must contact the network. Random access is the term for this technique, and mobile stations must understand how the random access channel is configured (preambles, time slots, etc.).

B. **Measuring Mobile Stations**
Inactive mobiles must perform mobility and perhaps positioning measures. The mobile stations use these measures to perform location update signaling with the network, which is required for paging to function.

C. **System Availability**
In theory, a mobile station may emit a probe and check for a network response to see if a system is available. It is, however, generally not deemed appropriate for mobile stations to transmit in any frequency range before they have received permission to do so (the mobile station may, e.g., be in a country where these frequencies are used for another service). As a result, inactive mobile stations must first detect the presence of a system before they can send anything.

In today's cellular systems, system overhead is communicated from each cell throughout the network. This is a relic from the first and second-generation voice-centric systems (Nordic Mobile Telephone (NMT), GSM). However, it is becoming a concern for the third and fourth-generation data-centric systems (HSPA, LTE). If system information distribution is thought of as a broadcasting problem, it can be solved more efficiently. Single-frequency network transmission technologies are the natural solution for providing broadcast services in OFDM-based systems, such as LTE [96].

Additionally, the fixed cells become increasingly difficult to justify with the rise of beamforming, MIMO, multicarrier, CoMP, multi-RAT, and reconfigurable antenna systems. This, as the argument goes, necessitates a rethinking of the concept of the cell. Researchers view the cell as something unique to one mobile station rather than something familiar to all mobile stations in a given area. A more efficient system operation can also be enabled by abandoning the idea that the coverage of the system information broadcast channel and the data channels should be identical. Figure 2.10 depicts this, with all mobile stations receiving system information via a Broadcast Channel (BCH).

The mobile station MS_1 is set up to accept data from a MIMO-capable cell that RBS_1 has made available. The matching signals from RBS_1, RBS_2, and RBS_3 are used by the mobile station MS_2 to communicate with a CoMP cell. The base stations in this instance are RBS_1, RBS_2, and RBS_3. A cell and the mobile station MS_3 interact using an omnidirectional

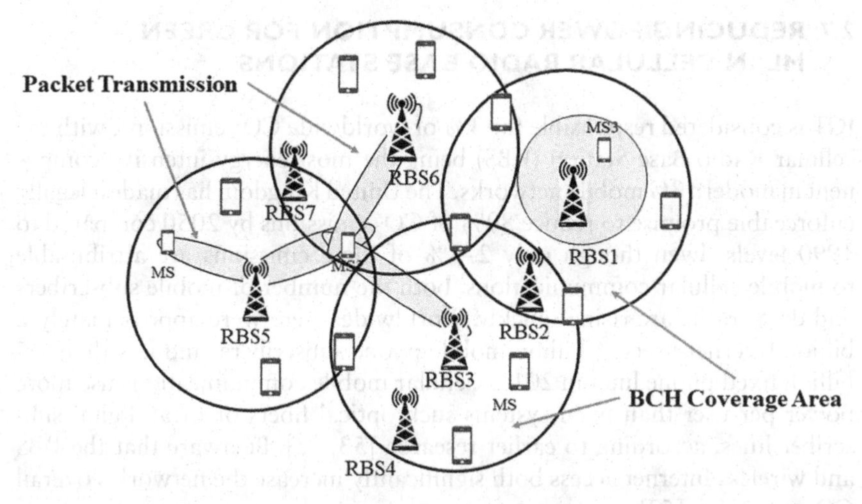

Figure 2.10 System operation enables separate coverage of system broadcast channels (BCH) and packed data transmissions.

The system information and transmission are not used by the base stations RBS_5, RBS_6, and RBS_7. The choice to configure a CoMP cell for MS_2 and a MIMO cell for MS_1 can then take into account the amount of traffic that MS_2 and MS_1 wish to communicate. Base station sleep mode can be effectively enabled by logically separating system information from data transmissions. The cells that are presented to active users can also be customized according to their requirements. This allows the system to effectively meet the first problem by flexible cells giving active users high packet data rates. Designing the overhead system broadcasts with the highest DTX both in time and in space successfully addresses the second difficulty [97].

2.6.3 Summary

The network power consumption in wireless communications is investigated in the part using a load-dependent power model that includes not only transmit power but also total grid power. The research revealed significant savings potential, particularly in low-load periods. We have talked about how to improve deployment, resource management, and network management. Depending on the application, each strategy can result in savings of up to 30%. The underlying hardware components should be improved further (not included in this section). These savings, however, cannot simply be added. If two or more techniques can cooperate or perhaps create synergy, that needs to be determined. On the other hand, different tactics can be used at various times of day or in different portions of the network such as dense urban areas versus rural areas, for example by network reconfiguration between the busy hour and night times.

2.7 REDUCING POWER CONSUMPTION FOR GREEN ML IN CELLULAR RADIO BASE STATIONS

ICT is considered responsible for 3% of worldwide CO_2 emissions, with the cellular Radio Base Station (RBS) being the most energy-intensive component in modern 3G mobile networks. The United Kingdom has made a legally enforceable promise to reduce 80% of CO_2 emissions by 2050 compared to 1990 levels. Even though only 2–3% of these emissions are attributable to mobile cellular communications, both the number of mobile subscribers and data traffic are rising quickly. Worldwide, there were approximately 2 billion Internet users, 4 billion mobile phone subscribers, and less than 1.5 billion fixed phone lines in 2011. Cellular mobile communications use more power per user than fixed systems such optical fibers or local digital subscriber lines, according to earlier research [53, 62]. Be aware that the RBS and wireless Internet access both significantly increase the network's overall CO_2 emissions [53].

Over the course of an RBS's 10- to 15-year lifespan, operating energy is the major source of cellular CO_2 emissions. Operators want equipment manufacturers to cut emissions as soon as possible. As an example, Orange wants to reduce energy consumption across Europe as well as reduce environmental impact with the goal of becoming net carbon zero by 2040. In a mobile device with a typical two-year lifespan, the manufacturing or embodied energy, also known as capital expenditure or CAPEX, makes up a significantly larger portion of the overall energy consumption [62]. According to a widely accepted estimate, 0.3 GW of electricity is used annually by the current 3.3 billion mobile phones [53] in use worldwide. The 3M global deployed cellular RBS uses 4.5 GW of electricity and is responsible for 2% of the world's CO_2 emissions. Embodied energy is a much smaller part of the RBS than operational energy use, as can be seen in Figure 2.10. More than 70% of the total energy is used by the radio access network [62]. Operator network energy costs occasionally even reach parity with human costs!

An archetypal 3G RBS with three 120-sector antennas consumes around 300–500 W of operational power from an AC source to produce 40 W of RF power. An average 3G RBS uses about 4.5 MWh of energy annually, which is already significantly less than the older GSM RBS models. With 12,000 base stations, the 3G mobile network in the UK uses more than 50 GWh annually. For instance, China Mobile serves its 500 million mobile clients in undeveloped regions by 500,000 GSM and 200,000 3G CDMA RBS. The initial iteration of the LTE standard, which was finalized at the end of 2008, provides a significant upgrade over third-generation (3G) wireless systems now in use. The RBS is transmitted to various users on different carrier frequencies using orthogonal frequency division multiple access (OFDMA).

Additionally, it permits the use of MIMO at both RBS and user equipment (UE) terminals, resulting in significantly higher data rates on each channel. The advanced high data rate LTE RBS has been estimated to use 3.7 kW of AC power for a cell with a 470 m radius [53]. However, more accurate industrial estimates range from 800 W to 1 kW, which is significantly more power than earlier 3G networks. RBS power usage per customer UE analysis reveals that LTE offers a significant energy advantage over 3G or HSPA. Over the past five to seven years, improving base station energy efficiency has taken on more significance. Mobile operators sought in 2006–2007 that the RBS energy use growth rate be changed from an annual increase of 8% to a more significant step-change increase in overall energy efficiency.

2.7.1 Energy conservation techniques in high-traffic-load environments

In the physically bigger current wireless cells, also known as macrocells, close to the RBS, the signal to interference noise ratio (SINR) is relatively high. This enables high data transmission rates in the LTE downlink, up to 5 bits/s/Hz, and supports high-bit-rate modulation techniques like 64-state Quadrature Amplitude Modulation (64-QAM). The feasible cell edge data rates are severely constrained by the SINR at the edge, which is significantly reduced as a result of signal propagation loss and local cell interference, which causes transmissions to use quadrature phase shift keying (QPSK).

The highest practical data rate will therefore be provided by smaller cells with a shorter distance to the cell edge. To guarantee a high overall network data throughput capability, an array of such small cells made up of femtocells or relay devices may be necessary. Although the smaller cells significantly boost energy efficiency [53], the requirement for a 2-D array of such small cells to maintain the same overall coverage offsets this. As a result, a single macro cell and a 2-D array of small cells will both yield essentially the same result. In terms of transmitted bits per Joule, the overall system energy efficiency is the same. The interference between the small cells in the array is not taken into account in this calculation.

The introduction of femtocells or Home NodeB devices has resulted in the ready availability of low-power cellular RBS access points for homes or small enterprises. At the same time, Pico cells may service densely populated urban areas. These methods are instrumental when it comes to servicing known traffic hotspots like offices or retail malls. These femtocells connect to the service network (through DSL or cable) and operate in the same RF bands as the cellular system. By December 2010, 18 operators had launched commercial femtocell services, according to Femto Forum [37], with another 30 committed to future deployment. With 12 million femtocells deployed by 2014, these are expected to increase at an even faster rate in future [38]. By 2020, it is expected that femtocell deployment will be even more widespread, but it will consume less than 5% of the total energy consumed by the radio access network.

Femtocell deployment aims to boost RBS effectiveness and achieve high bit/s/Hz/km2 data transmission over the network while consuming the least amount of energy possible. In order to preserve energy, the unused femtocells are shut off in the big cell concept, where a steadily rising proportion of clients are housed within the femtocells or Pico cells. For various user densities, this model replicates the relationship between the percentage of consumers who have femtocells and the amount of electricity used per user.

The system power consumption per user is used as an evaluation indicator, covering CAPEX and OPEX costs. However, when the number of femtocells deployed exceeds 60%, the system's energy usage increases again. This system power consumption accounts for the embodied CAPEX energy consumed in raw material extraction, transport, fabrication, assembly, and installation for the RBS and femtocell devices, as well as their final deconstruction. If there are only 30–60 users per macrocell, there is a reported gain in energy savings. The calculations depicted in Figure 2.11 resulted in a 3–40% overall energy savings. Researchers have studied the QoS performance of the macro cell-two-tiered femtocell's architecture [64], which is appealing for increasing the data throughput of users close to the cell edge. Two often used metrics are system capacity and coverage [50]. When designing the LTE system, system spectrum efficiency was viewed as a crucial metric for gauging anticipated performance [51].

2.7.2 Energy-saving various low-traffic loading methods environments

Because cellular networks are currently designed to handle peak traffic loads, sleep mode solutions [80] can lower energy usage by turning down all of an operator's RBS equipment or, in some cases, just a portion of it. Previous studies have found that energy savings of 8–29% are possible. Over 24 hours, cellular data traffic typically varies by 2:1, with the lowest load at 7 a.m. and the greatest at 9 p.m. [68]. They also find significant load changes among cells, with a further study revealing that only 40% of network cells carry 90% of data flow. As a result, further research into the utilization of sleep modes to promote better spectral resource sharing is necessary.

To reduce the energy used, switching off the radio transmitter whenever possible is a time-domain strategy [71]. Because the transmitter power amplifier often accounts for a large amount of the entire RBS energy budget, this can result in significant energy savings. When the traffic demand permitted, certain femtocells were put into sleep mode. There may be cost savings with other RBS components, such as transceiver hardware and baseband DSP (DSPs). Reduced heat production and energy use within the base station may also have a lessening effect on other base station components including cooling systems and power supply circuits. To maximize total Radio Frequency (RF) energy savings through time-domain sleep modes, the

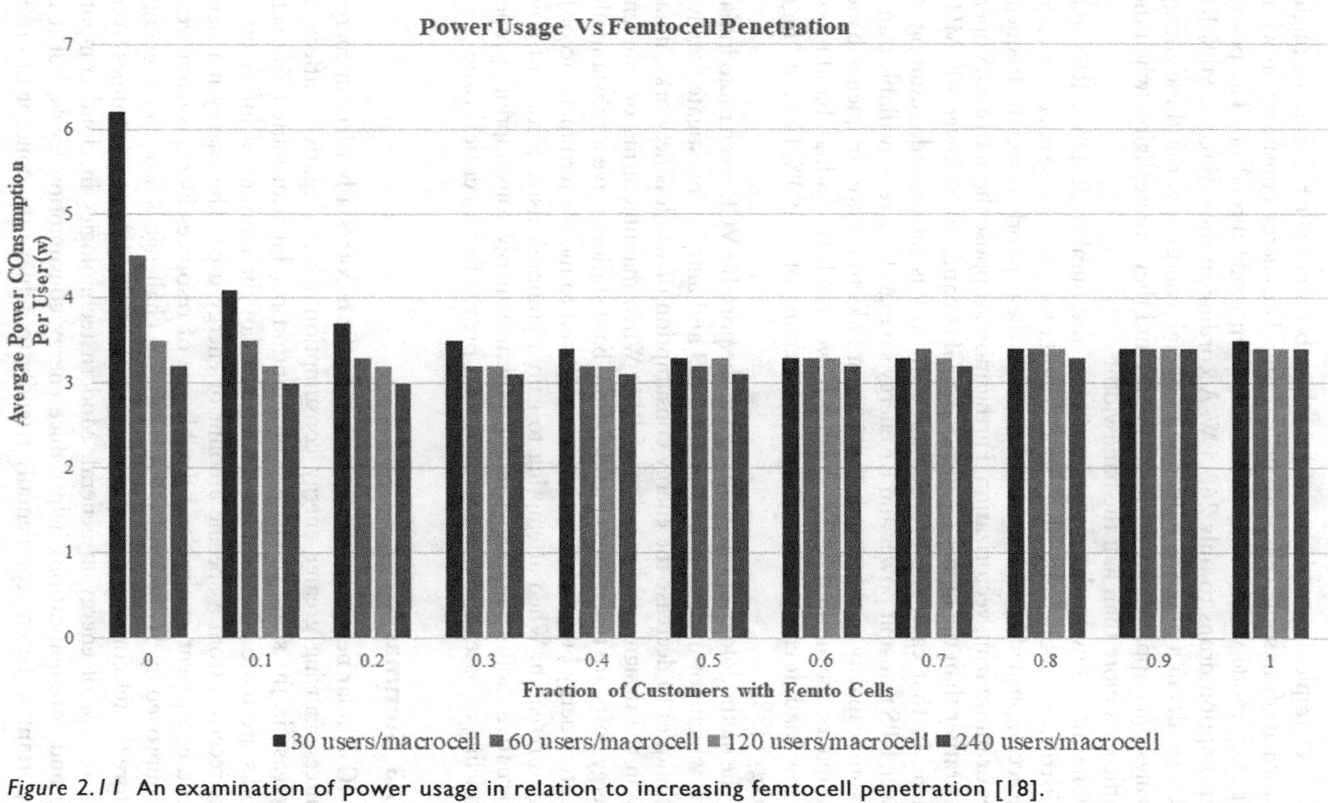

Figure 2.11 An examination of power usage in relation to increasing femtocell penetration [18].

optimal trade-off between control and data traffic power consumption must be identified.

For example, it was discovered that 9 sub-frames per radio frame might be turned off in some cells with little traffic. The energy consumption without sleep mode would be 600 W. With sleep mode turned on. The power consumption drops to only 262.5 W. According to the estimates, 90% RF ERG results in a 56.25% reduction in overall energy use. But these energy savings only apply when the load is light, and they are negligible when the traffic uses more than half the bandwidth!

One more low-load strategy minimizes the number of active RBS and redistributes mobile subscribers to RBS closer together. These energy savings are accomplished by turning off portions of the network at specific frequencies or minimizing vectorization. Furthermore, suppose the active RBS operates rather than the higher frequency 1.8 GHz band, in the lower 900 MHz range. In that case, the propagation path loss is minimized; resulting in lower, RBS transmit power and RF energy savings. If there is available transmission capacity in that band, reallocating links to a lower frequency band can reduce the amount of RF transmit power used. 10–15 high band users, or low load, it can produce a significant saving of 70–80% ERG at "high energy" [98].

The methodology of low-load used in the portable VCE green radio initiative, which got underway in January 2009 and aims to investigate creative techniques to decrease the energy consumption of wireless networks, notably in the design and operation of RBS. When manufacturing or inherent energy costs are taken into consideration, base stations have a substantially greater energy budget for operating compared to mobile terminals, according to research. When attempting to gain a comprehensive picture of how alternative radio technologies might minimize energy consumption, proper modeling of base station energy usage has proven to be a critical issue.

2.7.3 Summary

In 4G cellular networks, the large data traffic becomes a burden on energy sources, causing greater energy consumption which negatively affects a decrease in the share of energy consumption per bit transmitted. 5G networks are more efficient than 4G in terms of the amount of information bits received from a specific amount of energy used. The energy management, equipment quality, and flexible use of resources like spectrum were all improved by 5G networks. Currently, the challenge is how to use energy before the telecommunications industry, where the industry consumes from 2% to 3% of energy in general. Modern technologies in cloud computing and communications help reduce energy consumption levels, but it is important to be environmentally friendly to achieve the concept of green communications.

2.8 MACHINE LEARNING MANAGEMENT MODEL FOR GREEN WIRELESS NETWORKS

There has been little study into using ML to simulate green wireless networks for next-generation mobile situations. Nonetheless, some recent research [1, 25, 26] has focused on analyzing and leveraging data from the network and user surveys to model and improve the Quality of Experience (QoE). Figure 2.12 presents a framework that is one step ahead of the other efforts. The methodology for assessing or forecasting perceived quality of

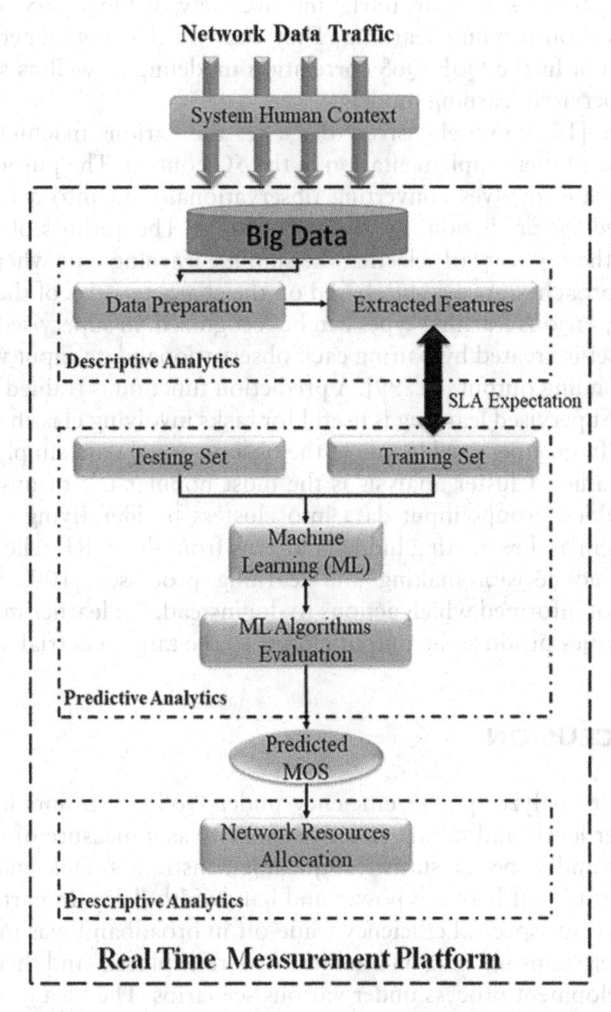

Figure 2.12 A framework for modeling perceived quality of experience using big data analytics.

experience is based on statistics gathered from the mobile network. In accordance with the suggested framework, this makes it possible for mobile network operators to control network performance effectively and provide users with an acceptable mobile Internet QoE. In this study, the state-of-the-art for modeling QoE with machine learning techniques is presented. It contains numerous trials in various scenarios, for various applications, and for various services [25].

Their research on the dimensions of QoE influence factors looks at many different scientific proposals. The authors concluded that while modeling QoE, three variables (person, system, and situation) should be considered. In [26], the authors gave an intriguing overview of QoE-QoS correlation models based on machine learning. They contend that non-supervised ML models do not fit the QoE-QoS correlation modeling as well as supervised or semi-supervised learning models.

Reference [10] comprehensively discusses the various machine learning techniques and their implementation in the 5G context. The purpose of ML is to anticipate involves converting observational data into a model that may be used for prediction, the output of input. The authors of this work emphasize the necessity of selecting the suitable learning type when employing ML approaches. As a result, based on the characteristics of the observational data, various learning types can be recognized. In supervised learning, a training set is created by pairing each observational data input value with its corresponding output [31, 99]. A prediction function is trained using this collection. Supervised learning is useful for tasks involving classification and regression. In unsupervised learning, the observational data simply contains the input values. Cluster analysis is the most popular use of unsupervised learning, which groups input data into clusters by identifying similarities between them and extracting hidden patterns from them. RL relies on iterative, dynamic decision-making and learning processes [100, 101]. The learner is not informed which actions to do; instead, the learner must decide which activities produce the output closest to the target via trial and error.

2.9 CONCLUSION

This chapter analyzes power efficiency under QoS constraints by looking at bit power levels and taking effective capacity as a measure of maximum throughput under specific statistical queuing constraints. This study characterized the trade-off between power and bandwidth-delay. In particular, the bit-power versus spectral efficiency trade-off in broadband was introduced, low-power systems under QoS limitations were explored, and then the analytical development process under various scenarios. The idea of the EE-SE trade-off was then introduced, along with an explanation of why it exists. The EE-SE trade-off for cellular systems based on BS collaboration and a practical PCM for CoMP systems was looked into. This realistic PCM was used to develop the EESE uplink swap for CoMP systems.

Following that, mobile operators are increasingly looking to improve the energy efficiency of their networks by enabling energy-saving modes. They are frequently introducing connectivity over heterogeneous networks, combining licensed (3G) and unlicensed (Wi-Fi) spectrums to achieve optimal economic capacity. The findings revealed the possibility of significant energy reductions of up to 50% in base station power. In addition, a load-based power model that incorporates not just transmission power but overall network power has been used to study network power usage in wireless communications. The study discovered significant cost-cutting opportunities, particularly in low-load scenarios. It explores how to improve deployment, resource management, and network management.

Depending on the use cases, each technology can result in savings of up to 30% of energy usage. Additional hardware component upgrades (not detailed in this chapter) are possible. These savings, on the other hand, cannot be added. It must be established if two or more technologies may coexist or even complement one another. On the other hand, different technologies can be employed at different times of the day, such as reconfiguring the network between busy hours and nighttime, or in different portions of the network, such as dense urban vs. rural areas. The results of evaluating the performance advantages of various technologies are presented in a number of ways that result in energy savings in circumstances with high and moderate traffic loads. The industry is leading the project in the hopes that good research findings would spur the development of energy-efficient wireless standards and devices in the future. Finally, using big data and machine learning approaches, a framework for building a global QoS management model for the future generation of the wireless ecosystem has been provided in this chapter. The centralized QoE strategy uses supervised ML approaches to find relevant KQIs for users, taking into consideration international standards. Unsupervised machine learning algorithms have been proposed to discover user influence variables and network performance abnormalities detect errors and optimize channel selection.

REFERENCES

[1] K. Lahiri, A. Raghunathan, S. Dey, and D. Panigrahi, "Battery-driven system design: A new frontier in low power design," in *Proceedings of ASP-DAC/VLSI Design 2002. 7th Asia and South Pacific Design Automation Conference and 15h International Conference on VLSI Design* IEEE, 2002, pp. 261–267.

[2] F. Fossati, S. Moretti, S. Rovedakis, and S. Secci, "Decentralization of 5G slice resource allocation," in *NOMS 2020–2020 IEEE/IFIP Network Operations and Management Symposium*, 2020, pp. 1–9.

[3] M. M. Saeed, R. A. Saeed and E. Saeid, "Survey of privacy of user identity in 5G: challenges and proposed solutions," *Saba Journal of Information Technology and Networking (SJITN)*, vol. 7.n. 1. Jul. 2019. ISSN:2312-4989.

[4] G. Ahmed et al., "Rigorous analysis and evaluation of specific absorption rate (SAR) for mobile multimedia healthcare," *IEEE Access*, vol. 6, pp. 29602–29610, 2018.

[5] Lana Ibrahim, Hana Osman, Abdalah Osman, Elmustafa Sayed Ali, "Impact of power consumption in sensors life time for wireless body area networks," *International Journal of Grid and Distributed Computing*, vol. 9, no. 8, pp. 97–108, 2016.

[6] X. Liu, X. Zhang, M. Jia, L. Fan, W. Lu, and X. Zhai, "5G-based green broadband communication system design with simultaneous wireless information and power transfer," *Physical Communication*, vol. 28, pp. 130–137, 2018.

[7] M. Saeed et al., "Preserving privacy of user identity based on pseudonym variable in 5G," *Computers, Materials & Continua*, vol. 70, no. 3, pp. 5551–5568, 2022.

[8] P. Jain and A. Gupta, "Energy-efficient adaptive sectorization for 5G green wireless communication systems," *IEEE Systems Journal* vol. 14, no. 2, pp. 2382–2391, 2019.

[9] Q. Wu, G. Li, W. Chen, D. Ng, and R. Schober, "An overview of sustainable green 5G networks," *IEEE Wireless Communications*, vol. 24, no. 4, pp. 72–80, 2017.

[10] G. Zhu, J. Zan, Y. Yang, and X. Qi, "A supervised learning based QoS assurance architecture for 5G networks," *IEEE Access*, vol. 7, pp. 43598–43606, 2019.

[11] K. Shafique, B. A. Khawaja, F. Sabir, S. Qazi, and M. Mustaqim, "Internet of things (IoT) for next-generation smart systems: A review of current challenges, future trends and prospects for emerging 5G-IoT scenarios," *IEEE Access*, vol. 8, pp. 23022–23040, 2020.

[12] R. A. Saeed, M. M. Saeed, R. A. Mokhtar, H. Alhumyani and S. Abdel-Khalek, "Pseudonym mutable based privacy for 5G user identity," *Computer Systems Science And Engineering*, vol. 39, no. 1, pp. 1–14, 2021.

[13] M. M. Saeed, R. A. Saeed, and E. Saeid, "Preserving privacy of paging procedure in 5 th G using identity-division multiplexing," in *2019 First International Conference of Intelligent Computing and Engineering (ICOICE)* IEEE, 2019, pp. 1–6.

[14] E. S. Ali, M. B. Hassan, and R. A. Saeed, *Chapter: Terahertz Communication Channel characteristics and measurements; Book: Next Generation Wireless Terahertz Communication Networks*. CRC Group, Taylor & Francis Group, USA, 2021.

[15] Y. Wang, J. Li, and H. Wang, "Cluster and cloud computing framework for scientific metrology in flow control," *Cluster Computing*, vol. 22, no. 1, pp. 1189–1198, 2019.

[16] H. Goumidi, Z. Aliouat, and S. Harous, "Vehicular cloud computing security: A survey," *Arabian Journal for Science and Engineering*, vol. 45, no. 4, pp. 2473–2499, 2020.

[17] T. Y. Fan and M. E. Rana, "Facilitating Role of Cloud Computing in Driving Big Data Emergence," in *2021 Third International Sustainability and Resilience* Conference: Climate Change IEEE, 2021, pp. 524–529.

[18] M. M. Saeed, R. A. Saeed, M. A. Azim, E. S. Ali, R. A. Mokhtar, and O. Khalifa, "Green Machine Learning Approach for QoS Improvement in Cellular Communications," in *2022 IEEE 2nd International Maghreb Meeting of the Conference on Sciences and Techniques of Automatic Control and Computer Engineering (MI-STA)*, Sabratha, Libya, 2022, pp. 523–528, doi:10.1109/MI-STA54861.2022.9837585.

[19] J. Kaur, M. A. Khan, M. Iftikhar, M. Imran, and Q. Haq, "Machine learning techniques for 5G and beyond," *Ad Hoc Networks*, vol. 9, pp. 23472–23488, 2021.

[20] M. Hassan, E. Ali, R. Mokhtar, R. Saeed, and B. Chaudhari, "NB-IoT: concepts, applications, and deployment challenges, book chapter (ch 6)," In *LPWAN Technologies for IoT and M2MApplications*, Elsevier, Berlin, Germany, 2020.

[21] S. N. Ghorpade, M. Zennaro, B. S. Chaudhari, R. A. Saeed, H. Alhumyani, and S. Abdel-Khalek, "Enhanced differential crossover and quantum particle swarm optimization for IoT applications," *IEEE Access*, vol. 9, pp. 93831–93846, 2021.

[22] J. Suomalainen, A. Juhola, S. Shahabuddin, A. Mämmelä, and I. Ahmad, "Machine learning threatens 5G security," *IEEE Access*, vol. 8, pp. 190822–190842, 2020.

[23] M. E. Morocho-Cayamcela, H. Lee, and W. Lim, "Machine learning for 5G/B5G mobile and wireless communications: Potential, limitations, and future directions," *IEEE Access*, vol. 7, pp. 137184–137206, 2019.

[24] R. S. Abdalla, S. A. Mahbub, R. A. Mokhtar, E. S. Ali, and R. A. Saeed, "IoE design principles and architecture," In *Internet of Energy for Smart Cities: Machine Learning Models and Techniques* CRC Group, Taylor & Francis Group, USA, 2021.

[25] A. Fehske, G. Fettweis, J. Malmodin, and G. Biczok, "The global footprint of mobile communications: The ecological and economic perspective," *IEEE Communications Magazine*, vol. 49, no. 8, pp. 55–62, 2011.

[26] I. Humar, X. Ge, L. Xiang, M. Jo, M. Chen, and J. Zhang, "Rethinking energy efficiency models of cellular networks with embodied energy," *IEEE Network*, vol. 25, no. 2, pp. 40–49, 2011.

[27] Mona Bakri Hassan, Elmustafa Sayed Ali, Rania A. Mokhtar, Rashid A. Saeed, and Bharat S. Chaudhari, "Chapter 6: NB-IoT: Concepts, applications, and deployment challenges," *LPWAN Technologies for IoT and M2M Applications*, First Edition; Academic Press Publishing; Elsevier, 2020.

[28] S. Bhaumik, G. Narlikar, S. Chattopadhyay, and S. Kanugovi, "Breathe to stay cool: adjusting cell sizes to reduce energy consumption," in *Proceedings of the first ACM SIGCOMM workshop on Green networking*, 2010, pp. 41–46.

[29] S. Ramanath, V. Kavitha, and E. Altman, "Open loop optimal control of base station activation for green networks," in *2011 International Symposium of Modeling and Optimization of Mobile, Ad Hoc, and Wireless Networks* IEEE, 2011, pp. 161–166.

[30] Mona Bakri Hassan, Mohammad Kamrul Hasan, Elmustafa Sayed Ali, Rashid A Saeed, Rania A Mokhtar, Othman O Khalifa, and Aisha Hassan Abdalla Hashim, "Performance evaluation of uplink shared channel for cooperative relay based narrow band internet of things network," in *2022 International Conference on Business Analytics for Technology and Security (ICBATS)*, IEEE, 2022.

[31] E. S. Ali et al., "Machine learning technologies for secure vehicular communication in internet of vehicles: recent advances and applications," *Security and Communication Networks*, vol. 2021, 2021.
[32] F. Mhiri, K. Sethom, R. Bouallegue, and C. Applications, "A survey on interference management techniques in femtocell self-organizing networks," *Journal of Network and Computer Applications*, vol. 36, no. 1, pp. 58–65, 2013.
[33] Ertshag Hamza, Honge Ren, Elmustafa Sayed, and Xiaolong Zhu, "Performance evaluation of queuing management algorithms in hybrid wireless ad-hoc network", *Springer Nature Singapore*, 2018.
[34] S. Cui, A. J. Goldsmith, and A. Bahai, "Energy-efficiency of MIMO and cooperative MIMO techniques in sensor networks," *IEEE Journal on Selected Areas in Communications*, vol. 22, no. 6, pp. 1089–1098, 2004.
[35] L. M. Correia et al., "Challenges and enabling technologies for energy aware mobile radio networks," *IEEE Communications Magazine*, vol. 48, no. 11, pp. 66–72, 2010.
[36] G. Auer et al., "How much energy is needed to run a wireless network?" *IEEE Wireless Communications*, vol. 18, no. 5, pp. 40–49, 2011.
[37] F. Héliot, M. A. Imran, and R. Tafazolli, "On the energy efficiency-spectral efficiency trade-off over the MIMO Rayleigh fading channel," *IEEE Transactions on Communications*, vol. 60, no. 5, pp. 1345–1356, 2012.
[38] Mona Bakri Hassan, Sameer Alsharif, Hesham Alhumyani, Elmustafa Sayed Ali, Rania A. Mokhtar, and Rashid A. Saeed. "An enhanced cooperative communication scheme for physical uplink shared channel in NB-IoT," *Wireless Personal Communication*, vol. 120, 2367–2386, 2021.
[39] S. Tombaz, P. Monti, K. Wang, A. Vastberg, M. Forzati, and J. Zander, "Impact of backhauling power consumption on the deployment of heterogeneous mobile networks," in *2011 IEEE Global Telecommunications Conference-GLOBECOM 2011*, 2011, pp. 1–5.
[40] A. Alnazir, R. A. Mokhtar, H. Alhumyani, E. S. Ali, R. A. Saeed, and S. Abdel-Khalek. "Quality of services based on intelligent IoT WLAN MAC protocol dynamic real-time applications in smart cities," *Computational Intelligence and Neuroscience*, vol. 2021, Article ID 2287531, 2021.
[41] H. Shin and J. H. Lee, "Closed-form formulas for ergodic capacity of MIMO Rayleigh fading channels," in *IEEE International Conference on Communications, 2003. ICC'03*, vol. 5, pp. 2996–3000, 2003.
[42] A. Afaq, N. Haider, M. Z. Baig, K. S. Khan, M. Imran, and I. Razzak, "Machine learning for 5G security: Architecture, recent advances, and challenges," *Ad Hoc Networks*, vol. 123, p. 102667, 2021.
[43] M. Dohler and H. Aghvami, "On the approximation of MIMO capacity," *IEEE Transactions on Wireless Communications*, vol. 4, no. 1, pp. 30–34, 2005.
[44] K. Son, E. Oh, and B. Krishnamachari, "Energy-aware hierarchical cell configuration: from deployment to operation," in *2011 IEEE conference on computer communications workshops (infocom wkshps)* IEEE, 2011, pp. 289–294.
[45] S. Zhou, A. J. Goldsmith, and Z. Niu, "On optimal relay placement and sleep control to improve energy efficiency in cellular networks," in *2011 IEEE International Conference on Communications (ICC)*, 2011, pp. 1–6.
[46] X. Wang, Y. Bao, X. Liu, and Z. Niu, "On the design of relay caching in cellular networks for energy efficiency," in *2011 IEEE Conference on Computer Communications Workshops (INFOCOM WKSHPS)*, 2011, pp. 259–264.

[47] R. Giuliano, C. Monti, and P. J. I. W. C. Loreti, "WiMAX fractional frequency reuse for rural environments," *IEEE Wireless Communications*, vol. 15, no. 3, pp. 60–65, 2008.
[48] D. Qiao, M. C. Gursoy, and S. Velipasalar, "The impact of QoS constraints on the energy efficiency of fixed-rate wireless transmissions," *IEEE Transactions on Wireless Communications*, vol. 8, no. 12, pp. 5957–5969, 2009.
[49] D. Qiao, M. C. Gursoy, and S. Velipasalar, "Energy efficiency of fixed-rate wireless transmissions under QoS constraints," in *2009 IEEE International Conference on Communications*, 2009, pp. 1–5.
[50] O. Arnold, F. Richter, G. Fettweis, and O. Blume, "Power consumption modeling of different base station types in heterogeneous cellular networks," in *2010 Future Network & Mobile Summit*, 2010, pp. 1–8.
[51] A. J. Fehske, P. Marsch, and G. P. Fettweis, "Bit per joule efficiency of cooperating base stations in cellular networks," in *2010 IEEE Globecom workshops*, 2010, pp. 1406–1411.
[52] M. B. Hassan, E. S. Ali, R. A. Saeed, "Chapter: Ultra-massive MIMO in THz communications," *Next Generation Wireless Terahertz Communication Networks*, CRC Group, Taylor & Francis Group, USA, 2021.
[53] J. Tang and X. Zhang, "Cross-layer modeling for quality of service guarantees over wireless links," *IEEE TRANSACTIONS on Wireless Communications*, vol. 6, no. 12, pp. 4504–4512, 2007.
[54] R. S. Alhumaima, R. K. Ahmed, H. Al-Raweshidy, and Networking, "Maximizing the energy efficiency of virtualized C-RAN via optimizing the number of virtual machines," *IEEE Transactions on Green Communications and Networking*, vol. 2, no. 4, pp. 992–1001, 2018.
[55] U. Challita, H. Ryden, and H. Tullberg, "When machine learning meets wireless cellular networks: Deployment, challenges, and applications," *IEEE Communications Magazine*, vol. 58, no. 6, pp. 12–18, 2020.
[56] K. Son, H. Kim, Y. Yi, and B. Krishnamachari, "Base station operation and user association mechanisms for energy-delay tradeoffs in green cellular networks," *IEEE Journal on Selected Areas in Communications*, vol. 29, no. 8, pp. 1525–1536, 2011.
[57] F. Heliot, M. A. Imran, and R. Tafazolli, "On the energy efficiency-spectral efficiency trade-off over the MIMO Rayleigh fading channel," *IEEE Transactions on Communications*, vol. 60, no. 5, pp. 1345–1356, 2012.
[58] F. Richter, A. J. Fehske, and G. P. Fettweis, "Energy efficiency aspects of base station deployment strategies for cellular networks," in *2009 IEEE 70th Vehicular Technology Conference Fall*, 2009, pp. 1–5.
[59] G. Fettweis and E. Zimmermann, "ICT energy consumption-trends and challenges," in *Proceedings of the 11th international symposium on wireless personal multimedia communications*, vol. 2, no. 4, p. 6, 2008.
[60] E. Oh, B. Krishnamachari, X. Liu, and Z. Niu, "Toward dynamic energy-efficient operation of cellular network infrastructure," *IEEE Communications Magazine*, vol. 49, no. 6, pp. 56–61, 2011.
[61] Reham Abdelrazek Ali, Elmustafa Sayed Ali, Rania A. Mokhtar & Rashid A. Saeed. "Blockchain for IoT-Based Cyber-Physical Systems (CPS): Applications and challenges," In De, D., Bhattacharyya, S., Rodrigues, J.J.P.C. (eds) *Blockchain based Internet of Things. Lecture Notes on Data Engineering and Communications Technologies*, vol 112. Springer, Singapore, 2022.

[62] S. J. Verdú, "Spectral efficiency in the wideband regime," *Transactions on Information Theory*, vol. 48, no. 6, pp. 1319–1343, 2002.
[63] A. Lozano, A. M. Tulino, and S. J. Verdú, "Multiple-antenna capacity in the low-power regime," *IEEE Transactions on Information Theory*, vol. 49, no. 10, pp. 2527–2544, 2003.
[64] O. Oyman and A. J. Paulraj, "Spectral efficiency of relay networks in the power limited regime," in *Proc. 42th Annual Allerton Conf. on Communication, Control and Computing*, Allerton, USA, 2004.
[65] J. Xu, L. Duan, and R. Zhang, "Cost-aware green cellular networks with energy and communication cooperation," *IEEE Communications Magazine*, vol. 53, no. 5, pp. 257–263, 2015.
[66] J. James and V. O. Li, "Base station switching problem for green cellular networks with social spider algorithm," in *2014 IEEE Congress on Evolutionary Computation (CEC)*, 2014, pp. 2338–2344.
[67] F. Yang et al., "Astar: Sustainable battery free energy harvesting for heterogeneous platforms and dynamic environments," in *Proceedings of the 2019 International Conference on Embedded Wireless Systems and Networks*, EWSN ACM, 2019, pp. 71–82.
[68] F. Meshkati, H. V. Poor, S. C. Schwartz, and N. Mandayam, "An energy-efficient approach to power control and receiver design in wireless data networks," *IEEE Transactions on Communications*, vol. 53, no. 11, pp. 1885–1894, 2005.
[69] R. G. Gallager, *Information theory and reliable communication.* (Vol. 588). Wiley, New York, 1968.
[70] S. Sampei, S. Komaki, and N. Morinaga, "Adaptive modulation/TDMA scheme for personal multimedia communication systems," in *1994 IEEE GLOBECOM. Communications: The Global Bridge*, vol. 2, pp. 989–993, 1994.
[71] Z. Chong and E. Jorswieck, "Analytical foundation for energy efficiency optimisation in cellular networks with elastic traffic," in *International Conference on Mobile Lightweight Wireless Systems*, 2011, Springer, pp. 18–29.
[72] F. Héliot, M. A. Imran, and R. Tafazolli, "Energy-efficiency based resource allocation for the scalar broadcast channel," in *2012 IEEE Wireless Communications and Networking Conference (WCNC)*, pp. 193–197, 2012.
[73] H. Kim and G. De Veciana, "Leveraging dynamic spare capacity in wireless systems to conserve mobile terminals' energy," *IEEE/ACM Transactions On Networking*, vol. 18, no. 3, pp. 802–815, 2009.
[74] Y. Sun, M. Peng, Y. Zhou, Y. Huang, S. Mao, and Tutorials, "Application of machine learning in wireless networks: Key techniques and open issues," *IEEE Communications Surveys & Tutorials*, vol. 21, no. 4, pp. 3072–3108, 2019.
[75] M. A. Marsan and M. Meo, "Energy efficient management of two cellular access networks," *ACM SIGMETRICS Performance Evaluation Review*, vol. 37, no. 4, pp. 69–73, 2010.
[76] K. Dufková, J.-Y. Le Boudec, M. Popović, M. Bjelica, R. Khalili, and L. Kencl, "Energy consumption comparison between macro-micro and public femto deployment in a plausible LTE network," in *Proceedings of the 2nd International Conference on Energy-Efficient Computing and Networking*, 2011, pp. 67–76.
[77] Z. Niu, Y. Wu, J. Gong, and Z. M. Yang, "Cell zooming for cost-efficient green cellular networks," *IEEE Communications Magazine*, vol. 48, no. 11, pp. 74–79, 2010.

[78] X. Weng, D. Cao, and Z. Niu, "Energy-efficient cellular network planning under insufficient cell zooming," in *2011 IEEE 73rd Vehicular Technology Conference (VTC Spring)*, 2011, pp. 1–5.
[79] K. Son, H. Kim, Y. Yi, B. Krishnamachari, Algorithms, and Applications, "Toward energy-efficient operation of base stations in cellular wireless networks," *Green Communications: Theoretical Fundamentals, Algorithms, and Applications*, 2012.
[80] G. Miao, N. Himayat, and G. Li, "Energy-efficient link adaptation in frequency-selective channels," *IEEE Transactions on Communications*, vol. 58, no. 2, pp. 545–554, 2010.
[81] N. Nurelmadina, M. K. Hasan, I. Memon, R. A. Saeed, K. A. Z. Ariffin, E.S. Ali, R. A. Mokhtar, S. Islam, E. Hossain, and M. A. Hassan. 2021. "A systematic review on cognitive radio in low power wide area network for industrial IoT applications," *Sustainability*, vol. 13, no. 1, p. 338, 2021.
[82] M. Hedayati, M. Amirijoo, P. Frenger, and J. Moe, "Reducing energy consumption through adaptation of number of active radio units," in *2011 IEEE 73rd Vehicular Technology Conference (VTC Spring)*, 2011, pp. 1–5.
[83] X. J. Yang, 3GPP R1-050507, TSG-RAN1, "Soft frequency reuse scheme for UTRAN LTE," Huawei, 3GPP R1-050507, TSG-RAN1, vol. 41, 2005.
[84] M. M. Saeed, M. K. Hasan, A. J. Obaid, R. A. Saeed, R. A. Mokhtar, E. S. Ali, Md Akhtaruzzaman, S. Amanlou, A. K. M., and Zakir Hossain, "A comprehensive review on the users' identity privacy for 5G networks," *IET Commun.*, vol. 16, pp. 384–399, 2022.
[85] Y. Qi, M. A. Imran, R. Tafazolli, and P. Wiley, "Energy-efficient mobile network design and planning," *Fog*, pp. 97–118, 2015.
[86] D. Zeller et al., "Sustainable wireless broadband access to the future Internet - The EARTH project," in *The Future Internet Assembly*, 2013, Springer, pp. 249–271.
[87] Y. Qi, M. A. Imran, and R. Tafazolli, "Energy-aware adaptive sectorisation in LTE systems," in *2011 IEEE 22nd International Symposium on Personal, Indoor and Mobile Radio Communications*, 2011, pp. 2402–2406.
[88] Nada M. Elfatih, Mohammad Kamrul Hasan, Zeinab Kamal, Deepa Gupta, Rashid A. Saeed, Elmustafa Sayed Ali, Md. Sarwar Hosain, "Internet of vehicle's resource management in 5G networks using AI technologies: Current status and trends," *IET Commun.*, vol. 16, pp. 400–420, (2022).
[89] A. A. Muthana and M. M. Saeed, "Analysis of user identity privacy in LTE and proposed solution," *International Journal of Computer Network & Information Security*, vol. 9, no. 1, 2017.
[90] Alaa M. Mukhtar, Rashid A. Saeed, Rania A. Mokhtar, Elmustafa Sayed Ali, and Hesham Alhumyani. "Performance evaluation of downlink coordinated multipoint joint transmission under heavy IoT traffic load," *Wireless Communications and Mobile Computing*, vol. 2022, Article ID 6837780, 2022.
[91] A. L. Stolyar and H. Viswanathan, "Self-organizing dynamic fractional frequency reuse in OFDMA systems," in *IEEE INFOCOM 2008-The 27th Conference on Computer Communications*, 2008, pp. 691–699.
[92] M. Kang and M. Alouini, "Capacity of MIMO Rician channels," *IEEE Transactions on Wireless Communications*, vol. 5, no. 1, pp. 112–122, 2006.
[93] E. Biglieri and G. Taricco, *Transmission and reception with multiple antennas: Theoretical foundations*. Now Publishers Inc., 2004.

[94] K. Seong, M. Mohseni, and J. M. Cioffi, "Optimal resource allocation for OFDMA downlink systems," in *2006 IEEE International Symposium on Information Theory*, 2006, pp. 1394–1398.
[95] Mamoon M. Saeed, Rashid A. Saeed, Rania A. Mokhtar, Hesham Alhumyani, and Elmustafa Sayed Ali. "A novel variable pseudonym scheme for preserving privacy user location in 5G networks," *Security and Communication Networks*, vol. 2022, Article ID 7487600, 2022 https://doi.org/10.1155/2022/7487600
[96] M. O. Mohammed, A. A. Elzaki, B. A. Babiker and O. Eid, "Spatial variability of outdoor exposure to radiofrequency radiation from mobile phone base stations, in Khartoum, Sudan," pp. 1–14, 2021.
[97] I. Gódor et al., "Green wireless access networks," *Green Communications*, pp. 475–517, 2012.
[98] Z. Ahmed, R. Saeed, S. Ghopade, A. Mukherjee, and M. M. A. Elsevier, "*Energy optimization in LPWANs by using heuristic techniques, LPWAN technologies for IoT and M2M applications, chap. 11*," 2020.
[99] R. S. Ahmed, E. Ahmed, and R. A. Saeed, "Machine Learning in Cyber-Physical Systems in Industry 4.0," in *Artificial Intelligence Paradigms for Smart Cyber-Physical Systems: IGI Global*, 2021, pp. 20–41.
[100] F. Alsolami, F. A. Alqurashi, M. K. Hasan, R. A. Saeed, S. Abdel-Khalek, and A. Ishak, "Development of self-synchronized drones' network using cluster-based swarm intelligence approach," *IEEE Access*, vol. 9, pp. 48010–48022, 2021.
[101] L. E. Alatabani, E. S. Ali, and R. A. Saeed, "Deep learning approaches for IoV applications and services," in *Intelligent Technologies for Internet of Vehicles: Springer*, 2021, pp. 253–291.

Chapter 3

Green federated learning-based models and protocols

Afaf Taik
Université de Sherbrooke, Shebrooke, Canada

Amine Abouaomar
Al Akhawayn University, Ifrane, Morocco

Soumaya Cherkaoui
Polytechnique Montreal, Montreal, Canada

3.1 INTRODUCTION

The increasing demand in smart applications and the boost of computational capacities of connected devices have enabled a wide range of applications at the edge such as image and speech recognition and recommendation systems. Such applications generate large volumes of data which will soon overwhelm the cloud infrastructure. While on-device computation becomes more efficient for model inference and simple data processing operations, there are still many efforts required to make machine learning (ML) models training at the edge possible.

Federated learning (FL) [1] is a promising distributed machine learning (ML) concept that allows training a shared global model on local datasets of multiple devices while preserving the private nature of the training data. In this setting, a centralized parameter server distributes an initialized learning model to participating devices. Each device trains the model on its local dataset and uploads the local update (i.e., model or gradient) to the parameter server. All received updates are then aggregated to obtain an updated global model which is sent to participating devices to perform a new local model training task starting a new iteration termed communication round. This iterative learning procedure is repeated until the global model converges [2].

In FL, repeated communication and extensive computation power required to train models, especially deep learning models, lead to high energy consumption. Optimizing the energy consumption throughout the learning process is of foremost importance, for an efficient on-device performance and reducing device drop-out due to battery drainage [3]. However, increasing the energy efficiency of FL requires rethinking the models' design and the overall learning process, while taking into account several challenges related to the wireless edge environment.

DOI: 10.1201/9781003230427-3

To begin with, heterogeneous and limited resources across clients pose new challenges to systems design as they limit clients' availability and ability to participate in FL [4]. Indeed, end devices have limited battery lives, varying energy arrival rates, and different transmission capacities and connectivity. As a result, only the subset of devices that have enough energy and good connectivity can be selected to locally train and upload the ML models [5]. Furthermore, both the size of the models that can be trained on-device and the number of local training iterations are limited by the devices' capacities [6].

Additionally, devices need to send full model updates during every communication round, which creates significant communication overhead [7]. ML models in general, and deep neural networks (DNNs) in particular, can have sizes in the range of gigabytes [8]. Such models require significant energy to be trained; thus, it may not be possible to train or deploy such models on resource-constrained devices. Moreover, the communication bandwidth is often limited or costly, and the devices have weak transmission power. Consequently, only a restrained number of devices can participate in the learning process.

Furthermore, the local datasets are characterized by statistical heterogeneity as they heavily depend on the local environment and client behavior [9]. The differences in data sizes and distributions usually cause several challenges to the learning process, especially the models' convergence. In particular, clients with large datasets will require more time and energy to train models compared to clients with less training samples [10].

Consequently, designing green FL protocols and training green models should consider the heterogeneity of both resources and data, and the constraints imposed by the wireless environment. Designing green FL protocols requires optimizing each step of the learning process, starting with model initialization. The design of energy-efficient FL models can be enabled through several research efforts in green machine learning that have been made over the past few years, such as lightweight embedded models (e.g., MobileNets [11]) and compression techniques. Compression techniques such as pruning [12] (i.e., removing less significant weights or connections from models) and quantization [13] (e.g., using float32 instead of float 64), can help reduce both the computation and the communication costs. Yet, such techniques are not directly transferable to the FL setting, due to the heterogeneity of the participants and the highly collaborative aspect of the process. Instead, these methods can be adapted and revisited to be more suited to FL.

Likewise, each step of the communication round should be tailored in an energy-efficient fashion. For instance, careful client selection and resource allocation are widely studied in literature [14–17] and are key to a green FL process. Given the novelty of FL, new optimization techniques have been proposed to adjust these steps. Recent works have proposed diverse approach for client selection such as giving priority to clients based on their channel states [18], or to clients with more important updates [19], or more diverse datasets [10]. Furthermore, compression techniques, coupled with judicious bandwidth allocation ratios, can be key to energy preservation during the upload of large models. Moreover, a more flexible choice of local training parameters (e.g., number of local epochs, number of used samples), based on the energy levels of the clients, is necessary to insure a broader and energy-efficient client participation. Such decision can, for instance, take into consideration the trade-off between computation and communication in FL, through choosing when to train and when to upload based on the dynamic cost of each operation.

In general, the goal of this chapter is to overview recent efforts toward green FL surrounding each of the steps in the iterative process and to discuss the required considerations and the possible future directions. The objectives of this chapter can be described as follows:

- Identify the different challenges facing the design of green FL models and protocols, alongside the different aspects to account for.
- Overview different techniques and enablers from green machine learning and highlight how they can be used as the initial model and enable green FL.
- Overview efforts toward green FL from the recent literature that try to optimize the different steps of the iterative process and discuss the opportunities they present and identify their shortcomings.
- Identify possible difficulties in current designs, and present upcoming challenges and future trends for green FL.

In this chapter, an overview of the considerations of green FL and key-enablers is presented. As illustrated in Figure 3.1, this chapter is organized as follows: This chapter starts in Section 3.2 by introducing FL as an iterative process and identifying the challenges surrounding energy efficiency in each of its steps. Then, after identifying key considerations for a green FL, several enabling hardware and software technologies are identified in Section 3.3. Techniques and methods for lightweight model design are presented in Section 3.4. Section 3.5 presents optimization methods for the different steps in the FL iterative process such as client selection, local training, models upload and aggregation, and personalization. Lastly, in Section 3.5, some future directions are presented, and conclusions are drawn, highlighting the importance of energy efficiency in the design of FL models and protocols.

3.2 FEDERATED LEARNING PROCESS AND KEY CONSIDERATIONS

FL was introduced in [7] as a solution to data privacy concerns. FL allows clients to collaboratively train a shared model while keeping their data on their devices and only sharing model parameters. FL is an iterative process heavily based on computation and communication trade-off. After the

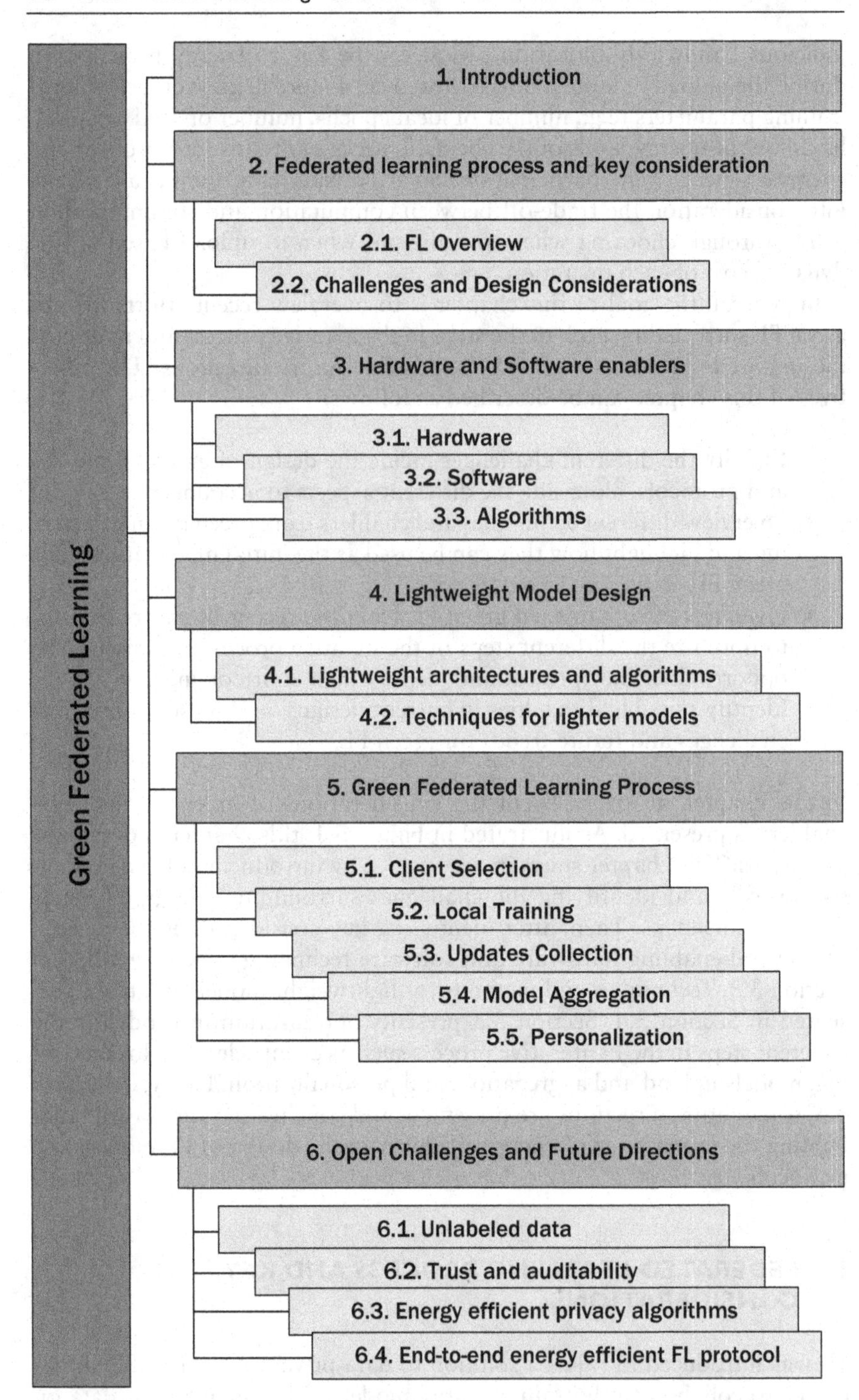

Figure 3.1 This chapter's organization.

model initialization at the centralized server, several iterations termed communication rounds will be repeated. In the following, we define a general template for FL and discuss the main challenges facing its adoption.

3.2.1 FL overview

FL can be used to train different models such as k-means, support vector machine (SVM), linear and logistic regression, and DNNs. The goal is to find the model parameters w that minimize the loss function f, which is used to calculate the error of the model on the training data. In supervised learning, a data sample i usually consists of two parts. The first part is the input (e.g., the pixels of an image, explanatory variables), which we denote as x_i, and the second part is the output (e.g., the label of the image, the predicted value of a regression problem) y_i is the desired output of the model. For some unsupervised models (e.g., k-means), the training data is composed of only an observation x_i. For each data sample i, we define the loss function as $f(w,x_i,y_i)$ for supervised models and $f(w,x_i)$ for unsupervised models. The loss function used for linear regression for instance is $f\left(w,x_i,y_i\right)=\frac{1}{2}\left\|y_i-w^T x_i\right\|^2$, where $\|\ .\ \|$ denotes the L_2 norm, and x^T is the transpose of x. Similarly, the loss function of k-means is $f\left(w,x_i\right)\frac{1}{2}\min_l\left\|x_{(l)}-w_{(l)}\right\|^2$ where $w\overset{\text{def}}{=}\left\{w_{(1)}^T,\ldots\right\}$ [20].

As illustrated in Figure 3.2, we consider a set of clients that collaboratively train a model with the orchestration of a server. A template for FL training can be considered as follows [21]:

1. *Global model initialization*: The global model architecture and training requirements are initialized in a centralized manner.
2. *Client Selection*: The server selects a subset of devices meeting training requirements. For instance, only devices that have enough battery, good reception and enough data samples could be selected for training.
3. *Global Model Broadcast*: The selected clients receive the global model alongside other training parameters (e.g., number of local epochs, optimizer) from the server.
4. *Local Training*: Each selected client locally trains the model on its local data. For each client j the loss function on the dataset D_j of data size $|D_j|$ is

$$f_j\left(w\right)=\frac{1}{\left|D_j\right|}\sum_{i\in D_j}f_i\left(w\right) \tag{3.1}$$

5. *Updates collection*: Each device uploads the local update to the server.

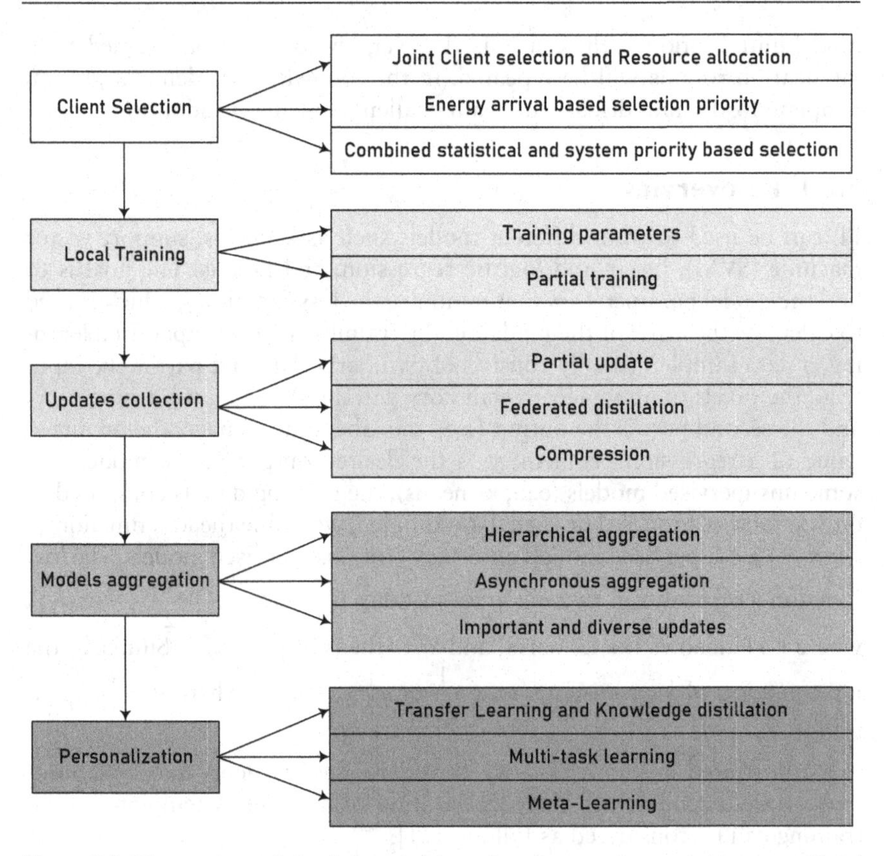

Figure 3.2 Illustration of the federated learning process for training an artificial neural network.

6. *Models aggregation*: The server updates the shared model based on the collected updates usually using a weighted average. For each client j the assigned weight can be for instance proportional to the size of its dataset such as in FedAvg [1], or to the freshness of its update in asynchronous FL [20]. We denote the assigned weight of the update of a client i with λ_i. The new global model is

$$w_g = \frac{1}{\sum_i \lambda_i} \lambda_i . w_i \tag{3.2}$$

7. *Personalization*: After the training following the FL process, each client personalizes the model either with local retraining or with a more optimized method, so it better suits the local data.

Each step should be carefully tailored for a green FL deployment. Having defined a general framework for FL, we next discuss the different challenges that need to be considered.

3.2.2 Challenges and design considerations

In contrast to other problems such as distributed learning, FL is more challenging as the number of participating devices is usually large, while the communication bandwidth and the available energy are fairly limited. We identify the following considerations for model and FL protocol design.

- **Resources heterogeneity:** Devices typically have different computation, storage, and communication capabilities. For example, the participants are equipped with different hardware (e.g., CPU, GPU, dedicated AI chips), network connectivity (e.g., 4G/5G, Wi-Fi), and energy source and level (e.g., plugged-in, unplugged) [22]. Such system heterogeneity magnifies the difficulty of some challenges such as fault tolerance and the straggler problem [23]. Indeed, the systems-related constraints on each device, and widespread heterogeneity, would often result in only a small fraction of the clients being available for training simultaneously. Moreover, the gaps in computational and communication resources create significant delays caused by stragglers in synchronous FL [24]. The developed algorithms should therefore be adaptive to the heterogeneous hardware throughout the FL process and tolerate partial participation. Furthermore, due to the high energy requirements of FL, green FL protocols will mostly depend on devices that can generate energy from renewable sources such as solar, kinetic, and ambient light. Such clients are characterized by intermittent energy availability, thus can only participate in the training process if they have generated enough energy, which imposes new considerations for the client selection process [25]. Indeed, the energy generation process is not uniform across the devices as some clients might generate energy in different periods of time and different frequencies.
- **Limited Resources:** The computing and storage resources of end devices are very limited and often shared among several applications, as it is often the case for smartphones and autonomous vehicles. As a result, the models should be carefully designed to take into consideration the communication and energy constraints through the choice of simpler and smaller models [11, 26, 27]. Furthermore, devices are frequently unavailable either due to low battery levels, or because their resources are used by other applications. As a result, device dropout is common during training and upload due to connectivity or energy issues. The communication resources, namely the bandwidth, are limited and commonly known to be FL's bottleneck. Communication-efficient methods

consisting of compressed [13] or partial model updates [28] are therefore key to preserving the energy and a better usage of the resources.

- **Statistical heterogeneity:** The datasets generated by the devices are highly heterogeneous, as they depend on the end user's behavior. In particular, this data generation paradigm violates the independent and identically distributed (i.i.d.) assumption and results in redundant datasets. Furthermore, the number of training samples varies significantly among devices, leading to different requirements for local training in terms of time and energy. As the goal of FL is to train one global model, the statistical heterogeneity of the data distributions often slows the model convergence and even leads to models that do not generalize to the entire population of devices. Consequently, several techniques such as multi-task FL, federated transfer learning, and local retraining were proposed to personalize the models. However, these solutions entail additional training costs.

3.3 SOFTWARE AND HARDWARE ENABLERS

Designing energy-efficient ML models has been widely studied, but the focus was often on inference, while work focusing on on-device model training is still in its infancy. Even so, research facilitating energy-efficient execution of ML algorithms can exploited and extended for FL in turn. The FL requirements in terms of computational and storage availability for on-device training might be higher than clients can provide. Current efforts toward on-device training and inference comprise dedicated hardware, specialized software, and lightweight model designs. Figure 3.3 summarizes the efforts discussed and explored in this section.

3.3.1 Hardware

Several research and industry efforts have been deployed to design dedicated hardware to enhance time and energy-efficient of ML algorithms on edge. The rise in popularity of ML is mainly due to the deployment of DNNs on graphic processing units (GPUs). GPUs are compatible with the parallelism of ML algorithms and highly suitable for matrix multiplication operations through the usage of multiple micro computing units. Several other hardware accelerators and dedicated chips were designed and exploited in the context of ML. Each hardware technology has some advantages over the others. For instance, GPUs have developer friendly tools, field-programmable gate arrays (FPGA) have high flexibility in terms of hardware reconfigurability but come at the price of complex hardware programming, and Application-Specific Integrated Circuits (ASICs) offer the highest computing efficiency but have long development cycles [29]. In the following, we present some of the ML enabling hardware and how they were optimized for better energy efficiency.

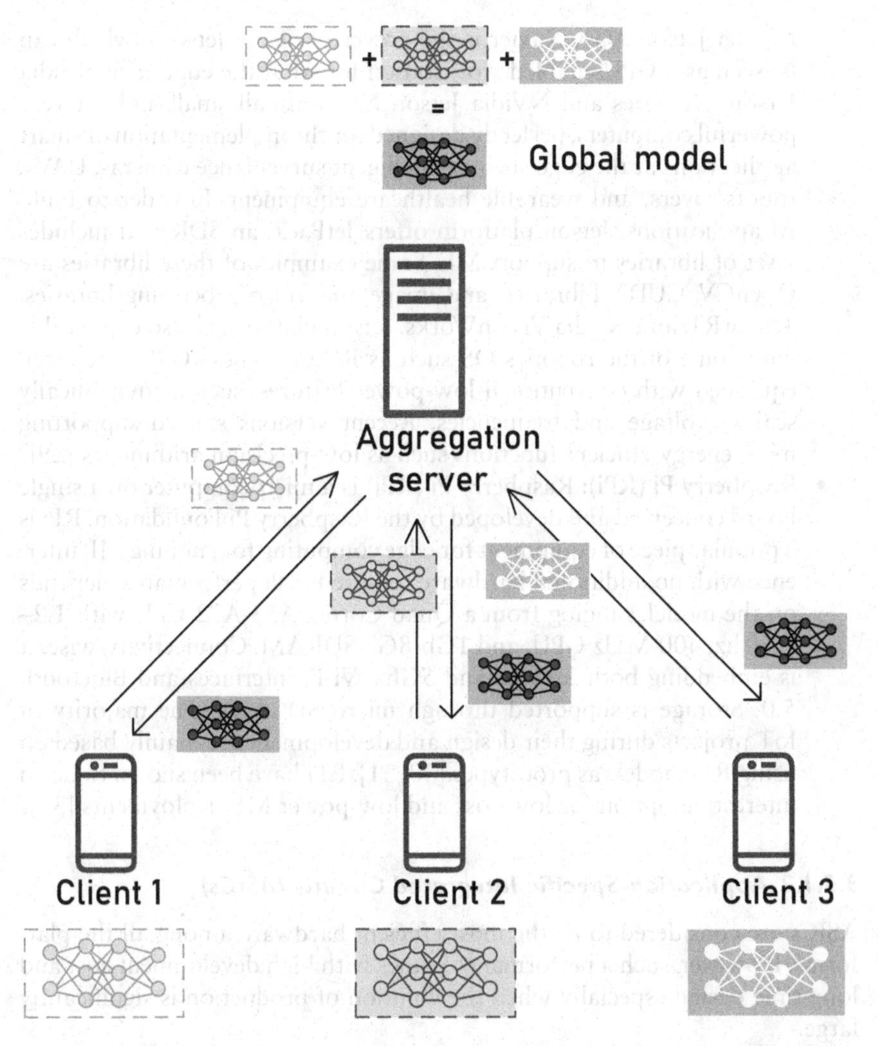

Figure 3.3 Green ML enablers in terms of hardware, software, and lightweight algorithms design.

3.3.1.1 General purpose embedded computation

GPUs have been widely used to accelerate ML both in cloud and edge; however, GPUs have high energy consumption. In an effort to overcome this concern, NVIDIA has released several embedded energy-efficient GPU products dedicated for autonomous vehicles (e.g., NVIDIA DRIVE PX2), robots, and drones (e.g., NVIDIA Jetson). Raspberry Pi is also a popular option for ML in edge environments owing to its low cost.

- **NVidia Jetson:** NVidia Jetson is a real-time data processing embedded computing. The Jetson family comprises, NVidia Jetson TX1, NVidia Jetson TX2, NVidia Jetson AGX Xavier, and the NVidia Jetson Nano.

NVidia Jetson AGX Xavier is the powerful of the Jetsons, which can be seen as a GPU-enabled workstation for AI at the edge. The NVidia Jetson TX series and NVidia Jetson Nano are all small and yet very powerful computers, perfectly designed for the implementation of smart applications at the edge such as intelligent surveillance cameras, UAVs, robots/rovers, and wearable healthcare equipment. In order to build AI applications, Jetson platform offers JetPack, an SDK that includes a set of libraries to support ML. Some examples of these libraries are OpenCV, CUDA Libraries, and image and video processing libraries, TensorRT, and Nvidia VisionWorks. Jetson platform is also compatible with some of the robotics OS such as ROS37. These GPUs are often equipped with conventional low-power features, such as dynamically scaling voltage and frequencies. Recent versions started supporting more energy-efficient functions such as low-precision arithmetics [29].

- **Raspberry Pi (RPi):** Raspberry Pi (RPi) is a microcomputer on a single board conceived and developed by the Raspberry Pi Foundation. RPi is a popular piece of equipment for edge computing for running ML inference with no additional hardware required. RPi performance depends on the model, ranging from a Quad Cortex A53-A72 CPU with 1.2–1.5Ghz, 400 MHz GPU, and 1Gb-8Gb SDRAM. Connectivity-wise, it is embedding both 2.4Ghz and 5Ghz Wi-Fi interfaces and Bluetooth 5.0. Storage is supported through micro SD cards. The majority of IoT projects during their design and development are mainly based on using RPi models as prototypes [30, 31]. RPi have been shown to be an interesting option for low-cost and low-power ML deployments [32].

3.3.1.2 Application-Specific Integrated Circuits (ASICs)

ASICs are considered to be the most efficient hardware among all the platforms. However, such a performance comes with high development cost and long-time design especially when the volution of production is significantly large.

- **Tensor Processing Units (TPUs):** are introduced to increase the performance of neural networks. TPUs provide up to 30 times the performance than a CPU and up to 80 times the performance per watt of existing GPUs. Such benefits enable numerous applications and hardware devices (such as the Raspberry Pi) to run advanced neural networks at an affordable cost and with minimal performance overhead. An important factor in the superior performance that TPUs deliver compared to CPU/GPU acceleration is the quantization process and systolic arrays [33], which contribute to a reduced memory footprint and lower power consumption. A systolic array supports matrix operation and allows neural network computation to be represented as matrix operations.

TPUs are not the only option adopting a systolic array-based computation; other options include HyGCN [34], Caffeine [35], and Tensaurus [36].

- **Coral USB Accelerator:** a coprocessor to assist ML inference on small devices such as Raspberry Pi. Coral is compatible with Linux-based OS and connected to the main system through a USB port. Compared to other rivals, Coral is 10 times faster than the Movidius, 3 times faster than the Nvidia Jetson Nano, and 3 times faster than the Raspberry Pi when running a MobileNetV2 [6] single-shot detector ML model. Additionally, it provides least energy consumption in both idle and peak utilization [37]. It is worth mentioning the Coral Dev Board by Google that leverages Edge TPUs as a co-processor to perform ML operations. The first component of the Coral Dev Board is the baseboard that have 40-pin GPIO header to attach different sensors or actuators for IoT devices. The other component is the system-on-module having a Cortex-A53 processor additionally to a Cortex-M4 core, 1GB of RAM and 8GB of memory helping running Linux on the edge TPU.
- **Movidius by Intel:** is neural compute stick is as simple as a pluggable USB device for a vision processing unit. This accelerator provides DNN inference in resource-restrained ordinary edge devices such as cameras. This chip is energy-efficient and able to run 100 Gigaflops using only 1 watt of energy. For instance, Movidius can be installed within robot computational units and being placed in a sinister area, and adequately take a decision regarding saving lives. Movidius provides unmanned vehicles (aerial or terrestrial) the ability to acquire intelligence through a variety of API and development kits. Movidius is cross-platform and independent on the edge device running OS, such as Debian, BSD, Windows, and Raspberry. Finally, software-wise, Movidius supports a plethora of ML frameworks, namely, Caffe, Apache MXNet, Pytorch, and TensorFlow.

3.3.1.3 FPGAs

The success of ARM processors is reflected on its existence and apparition on a plethora of edge devices such as smartphones, virtual reality equipment, robotics platforms, and medical instruments. ARM processors are suitable for ML application and have not only enabled the acceleration of ML algorithms and inference performance on edge devices, but also made them energy-efficient [38–40]. The architecture of such a system consists of an ARM NN 34 and the ARM compute library. ARM NN 34 represents the engine of inference which plays the role of a bridge to decrease as much as possible the gap between legacy Neural Network frameworks and the ARM ML processor. ARM compute library, on the other hand, is a library offering a set of functions dedicated to the manipulation of the ARM processors. Additionally, ARM ML supports other high-level frameworks such as

TensorFlow Lite, ONNX, and Caffe. ARM processors can offer even better performance by efficiently designing the neural network kernels [41].

3.3.2 Neuromorphic processors

Neuromorphic processors have architectures that are highly compatible with complex neural network algorithms. In contrast with Von Neumann machines, these brain-inspired processors bring the memory and the computational elements closer as means to efficiently implement ML algorithms. Neuromorphic processors use silicon technology to engineer neurons and synapses that can be programmed to emulate the human brain in carrying out tasks [42]. They can leverage the synaptic plasticity capabilities for online training and inference, using local information, which makes such hardware particularly suitable for problems requiring fast adaptation to new data [43]. Such architecture only consumes milliwatts-scale energy when processing artificial neural networks (ANNs) [44], making it an attractive solution for green ML. Existing examples of neuromorphic processors include Loihi by Intel [45] and IBM TrueNorth [46].

3.3.3 Software

The ML process requires a whole set of different tools to train the models. This ranges from the careful choice of software for the lower layers (i.e., the operating system and embedded hardware managers) to higher layers involving programming languages and frameworks. The choice of appropriate ML framework has a direct influence on energy consumption of the end devices. The smaller the hardware used, the lighter the ML packages should be. Therefore, deploying an ordinary operating system (OS) (i.e., used for standard computers and equipment) on a resource-constrained device (e.g., the raspberry range of microcomputers) would increase its power consumption even before reaching the stages of model training or implementation of an ML-based approach. A variety of lightweight platforms can be used to lower resource demands, such as TinyOS [47], a component-oriented embedded operating system for low-power devices. Another example is the Phi-Stack [48], a hardware-software stack jointly designed to inherently handle both intelligent and web-based applications at reduced cost, memory, and power consumption. Furthermore, in the context of on-device ML and FL, several efforts were deployed toward packages, libraries, and frameworks. Additionally, energy-aware development tools and power consumption evaluation in ML are essential for guiding the development of green FL [49]. An evaluation of how some ML packages perform on-device in terms of energy can be found in [50]. One of the key findings though is that none of the packages were able to perform the better in all the experiments, indicating that there is a large room for improving the performance of ML packages on the edge. In what follows, we focus on some of the existing projects that enable the evaluation of ML, FL, and energy models on devices.

3.3.3.1 On-device ML

- **Tensorflow-related projects:** Tensorflow is one of the largely adopted DL libraries especially in production, owing to its easy usage through Keras [51] interface. Numerous projects and libraries were created, extending its functionalities. Tensor/IO [52] is a platform that brings the power of TensorFlow to mobile devices such as Android and iOS by enabling its interaction with file systems and formats that are not built-in in Tensorflow. This platform eases the process of implementing and deploying Tensorflow models on mobile phones through a set of choices of back-end programming languages, such as Objective-C, Swift, Kotlin, and Java. TensorFlow Lite (TFL) [53] is an open-source DL framework for edge-aware inference. It addresses some of the performance key constraints such as privacy, connectivity, latency, size, and, most importantly, the power consumption of the edge devices. TFL is cross-platform OS-wise and capable of running over Android, iOS, embedded Linux. By way of performance illustration, TFL's binary is barely 1Mbits if we consider a 32bit ARM-based hardware, dropping to as low as 300Kbits binaries for classification use cases. While TFL is only developed for inference, the same tools used for TFL constitute a promising basis for energy-efficient on-device training. Furthermore, differential privacy is enabled in Tensorflow through its differential privacy project Tensorflow Privacy [54].
- **PyTorch Mobile:** PyTorch Mobile [55] Dedicated for smartphones, it is part of the PyTorch environment designed to support all ML phases from training models to deploying it on the target platforms. Various APIs are provided to handle ML in mobile applications. It may support TorchScript IR scripting and tracing. It can also support many of the ML accelerators such as GPUs, FPGAs, and TPUs. The ease of optimization for mobile deployment is enabled through the mobile interpreter. It currently supports a set of use cases such as object detection, image and video processing, speech recognition, and interacting chatbots.
- **uTensor:** uTensor [56] provides a free embedded learning development framework that assists in prototyping and fast IoT-Edge deployment (uTensor, 2021). It features an inferencing core, a graph processing tool, and a data collection architecture. It basically took a neural network model with Keras for the purpose of training. Then it converts the trained model into a C++ model. uTensor assists in converting a model for proper deployment in Mbed, ST, and K64 chips. uTensor is a very small module requiring only 2 KB on of disk storage. Python SDK is employed for customizing the uTensor from scratch, depending

on uTensor client interface, or python-based tools such as Jupyter. Initially, a model is built, quantized, and then adapted for the targeted devices.
- **μTVM:** μTVM [57] MicroTVM for tensor virtual machines (TVM) are an extension to facilitate the execution of tensor-based programs on different chips. It allows the optimization of TVM programs through the AutoTVM platform (μTVM, 2021). The workflow in uTVM is initialized by connecting the microcontroller with a high-end hardware through USB-JTAG port (computer, or cloud servers) capable of running the TVM in background. The connection between the microcontroller and the desktop is provided by OpenOCD executed on the high-end hardware which leverages device agnostic TCP port. Users in return should be providing a set of tools such as C cross-compiler suites for microcontrollers as well as, methods to read, write, and execute on the memory of the device, in addition to code segment to prepare the device to execute different functions. Finally, binaries are loaded into the device after being compiled.

3.3.3.2 Federated learning libraries

Open-source libraries and platforms facilitating FL are still under development. Current efforts are mostly aimed at enhancing privacy and security or enabling more research and experimentation.

- **PySyft:** PySyft [58] is an open-source library allowing private and secure ML through the extension of legacy deep learning frameworks in a lightweight, transparent, user-friendly way [58]. Its objective is to promote privacy mechanisms in ML by easing access to these techniques through Python as the popular tool in the research and data science communities. In addition, PySyft is aiming to be extensible so that security and privacy methods, namely differential privacy, multiparty computation, and homomorphic encryption, can be implemented and integrated in a smooth and flexible way, in both desktop and mobile environments. PySyft functions only in simulation and experiment mode but can be extended to support FL in production through the integration with other projects such as PyGrid.
- **FATE (Federated AI Technology Enabler):** FATE [59] is an open-source project for secure FL initiated by the Webank AI division. It enables similar security mechanisms as PySyft but is only focused on server/desktop deployments. It delivers a host of FL algorithms, such as deep learning, logistic regression, tree-based algorithms, and transfer learning.
- **Paddle Federated Learning Framework (PFL):** PFL [60] is an open-source FL framework that uses PaddlePaddle, a DL platform, for NN implementation, and also offers linear regression. Similarly to other frameworks, it implements differential privacy and secure aggregation. It supports both simulated and federated nodes using Docker

containers. PFL has high resource requirements (around 6 GB RAM and 100 GB of storage), which limit the usage of PFL in IoT systems.

- **TensorFlow Federated (TFF):** TFF [61] an open-source framework for ML and other decentralized data computations [62]. TensorFlow Federated provides an FL API, which is a high-level API implementing FL on a single machine through treating different files as separate clients, and Federated Core API which provides a low-level framework underlying the FL API to execute and simulate user-defined computational types and control over the orchestrations. TFF is currently aimed at research and experimentation, but it can be key for simulations before production-level deployments.
- **Flower:** Flower [63] is model framework agnostic as it is compatible with most existing ML frameworks, including scikit-learn, Pytorch, and Keras. It is compatible with a plethora of servers and devices, as it works on mobile (e.g., Android, iOS), cloud platforms (e.g., AWS, GCP), and different hardware (e.g., Raspberry Pi, Nvidia Jetson).

3.3.3.3 Energy and power consumption evaluation

Energy consumption has been largely studied through a wide range of computing software and hardware. However, the adoption of energy as a performance indicator did not receive much attention as most research is primarily focusing primarily on getting increased levels of accuracy. Furthermore, most of ML research comes with the premise of abundance of computation resources, thus deploying large models in datacenters and training for extensive periods of time. One of the leading causes of overlooking energy consumption in ML research is the lack of unified approaches to evaluate energy consumption. It was suggested to exploit and extend energy estimation approaches from the computer architecture community to estimate the energy consumption of ML algorithms [49]. However, as energy consumption depends also on the used hardware, creating general purpose estimators becomes challenging. As a result, studies and tools for different hardware architectures have emerged, focusing for example on GPUs [64, 65] or mobile devices [66].

Some examples of packages and tools that can help developers optimize their code in a green manner include CodeCarbon [67],[1] a package for aimed at Python code and estimating the amount of carbon dioxide (CO2) produced by its execution. Another interesting project is JEPO [68], a plugin for Eclipse designed and developed to offer a set of suggestions and recommendations for software developers to energy-wise enhance the efficiency of their ML code in both real-time writing and refactoring code. Other frameworks were proposed for automating these changes such as Chameleon [69]. Another interesting project for green FL is DeLight [70], which focuses on modeling the simple feed-forward neural networks energy consumption at the training stage. The design covers energy efficiency for basic arithmetic operations and communication cost in a distributed training configuration. Such work can be extended to other types of ANNs for energy estimation in FL settings.

3.4 LIGHTWEIGHT MODEL DESIGN

Many techniques were previously proposed in ML for green models' design. The proposed methods in ML design focus on reducing the memory and computation requirements where several modifications and tweaks were proposed to make these models more resource efficient. This is often achieved through keeping the models small by using as few parameters as possible as well as minimizing the number of operations and memory accesses.

3.4.1 Lightweight architectures and algorithms

Several lightweight model architectures and algorithms were proposed as alternatives to more computationally heavy standard models. In particular, these consist of variants of neural network architectures, and optimized implementations of several models such as decision trees.

- **Convolutional neural networks (CNNs):** It is often challenging to deploy CNNs on edge-devices due to limited memory and power, albeit it is very desirable given the wide applications they enable. As a result, several CNN architectures were developed. Examples include MobileNets [11], a class of network architectures for mobile and embedded applications, allowing to choose a small network adapted to the resource restrictions of the device and application. MobileNets are based on a streamlined architecture that uses depthwise separable convolutions, which consist of decomposing convolution filters into two simpler operations to build light-weight CNNs. In the case of a convolutional filter of size K×K, with an input of M channels and an output of N channels, the depthwise separable convolution reduces the computational cost by a factor of $1/N+1/K^2$ and the number of parameters by a factor of $1/N+M/(K^2 \times M+1)$ compared to a standard convolution. In addition to depth wise separable convolutions, ShuffleNet [71] uses two operations called group convolutions and channel shuffling for lighter and faster models without much of an impact on accuracy. Some other examples include SqueezeNet [72] which relies on reducing the sample size using special 1×1 convolution filters while still achieving state-of-the-art accuracy, YOLO [73], and single-shot detector [74], which predict both the location and class of the object at the same time.
- **Recurrent neural networks (RNNs):** RNNs are vastly used for time-series and natural-language processing applications. For reduced model size, Coupled Input and Forget Gate (CIFG) cells and Gated Recurrent Units (GRU) are often used instead of standard Long Short Term Memory (LSTM) cells and show comparable results with fewer parameters [75]. For example, Tensorflow Lite uses CIFG as the default recurrent neural network cell. Recently, newer architectures that can be trained and executed on highly constrained devices such as Arduino Uno were proposed. FastRNN and FastGRNN [76] were

proposed as two small size architectures that provide stable training and good accuracy. FastRNN is built through adding a residual connection with two additional parameters for a more stable training. FastGRNN is a gated version of FastRNN that does not increase the number of parameters by quantizing the weights to integer values. Overall, FastGRNN has up to four times fewer parameters than other gated RNN cells such as LSTM without affecting the accuracy.

- **Spiking neural networks:** Spiking Neural Networks (SNNs) are an energy-efficient alternative to the traditional ANNs and use a more brain-like model. Significant progress has been made in the multi-layer training of supervised and unsupervised deep SNNs. The efficacy of SNNs was demonstrated in image classification tasks, while work on training recurrent SNNs is still in its infancy, as the most advanced work on recurrent SNNs are the liquid state machines [77]. Furthermore, the simulation of SNNs on traditional computers is time-consuming because of the time discretization used in spikes, which slows down their large-scale adoption. SNN hardware implementation, especially on neuromorphic accelerators, may facilitate the use of SNNs as a means to reduce power consumption by increasing their evaluation speed. Currently, very few efforts were made to deploy and evaluate the potential of FL with SNN [43, 78].
- **Support Vector Machines (SVMs):** SVMs are widely used for classification, regression, and outliers detection. Their efficiency in such problems made their training and inference important in edge networks. Accordingly, Edge2Train [79] was developed to train SVMs on a variety of open-source small micro-controller units that have limited Flash, SRAM, and no floating-point support. Edge2Train is implemented in C++ and compared with SVM's implementation on CPU Python's library scikit-learn; it was shown to have less energy consumption across several tasks.
- **Decision Trees:** Bonsai [80] is a novel lightweight tree learner. Its idea is based on learning a single, shallow, sparse tree that results in a reduced model size, through limiting it to a few nodes that make non-linear predictions. Additionally, all nodes of Bonsai are learnt jointly rather than one by one. Furthermore, Bonsai learns a sparse matrix which projects all data points into a low-dimensional space, allowing the model to fit in a few Kilobits of flash.

3.4.2 Techniques for lighter models

Techniques to obtain lighter models to train include Transfer learning, knowledge distillation, and network architecture search.

- **Transfer learning and FedPer:** Careful model design can also use transfer learning, an ML method where a model developed for a certain task is reused as the starting point for a new model for another similar task [81]. Transfer learning is widely used in computer vision and natural language processing, as these are computationally heavy tasks. For FL, a base network can be trained on a publicly available dataset, or a publicly available model can be used for this purpose. Then the learned features can be repurposed, i.e., transferred to the global model to be trained on the federated datasets. This process is more likely to work efficiently with more common and general features (i.e., suitable to both base and new tasks) instead of specific to the base task. This can be achieved, for example, by using the first layers of a pre-trained CNN of an image classification application. The first layers of the CNN extract general features (i.e., shapes and contours) for images. Because these features need to be extracted regardless of the image classification task, these layers can be reused in different applications and thus reducing the training cost. Transfer learning can be used to choose the initial model for training and focus on training the last models in a federated manner. As the convolutional layers of a CNN are the most energy-consuming ones, transfer learning can help reduce the costs significantly.

To overcome problems related to statistical heterogeneity, FedPer [82] was proposed. FedPer is a neural network architecture where the first layers (called base layers) are trained following the standard FL protocol, and the final layers are trained locally for each client using a variant of gradient descent. The difference between FedPer and transfer learning is that while in the latter, all the layers are first trained on another task or on publicly available data, and then all or some layers are retrained on local data, for FedPer, the training starts with the base layers on global data using FL, and then the last layers are trained on local data only.

- **Knowledge Distillation:** Knowledge Distillation involves creating a smaller neural network (i.e., student network) that imitates the behavior of a larger one (i.e., teacher network). This is achieved by using the prediction outputs of the more powerful model to help train the smaller one. The student model can then be deployed on resource constrained devices and still achieves considerable accuracy levels. It was also observed that classifiers can be trained faster and more accurately using the outputs of other classifiers as soft labels, instead of from ground truth data. Knowledge distillation was used in FL to adapt to devices heterogeneity. In this setting, each client owns not only their private data, but also uniquely designed models [83].
- **Network Architecture Search (NAS):** NAS is widely used for DNNs to automatically tune the architecture and its corresponding hyperparameters. However, searching for optimized neural architectures in FL settings is particularly challenging due to the system and data heterogeneity. Traditional approaches for NAS include reinforcement learning, evolutionary algorithms (e.g., genetic algorithm), and gradient descent. Nonetheless, such methods are not suitable for resource constrained devices. Recent efforts toward federated NAS include offline and online NAS methods [84].

For example, DecNAS is an offline federated NAS framework [85] that uses three key techniques: parallel model tuning, dynamic training, and candidates' early dropping. The DecNAS starts with an initial global model (e.g., Mobilenets-based model) and a list of resource constraints in the server and uses pruning to create a list of candidate global models, then distributes to different subsets of clients. During each communication round, each subset of clients trains the received group models for a number of predefined epochs and tests them on local validation datasets. Then, clients upload both local test accuracies and validation dataset sizes to the server, where new models are aggregated and a weighted accuracy for each model is calculated. The models that perform poorly are eliminated and a new round begins until a satisfactory global model is obtained. However, this approach is not resource-friendly as training candidate models consume additional communication and computation resources. Consequently, online federated NAS frameworks were developed [86–88].

Online federated NAS jointly trains and optimizes the neural architecture. The common idea of several research efforts is to start with a global model called Supernet [89]. The Supernet is based on repeated directed acyclic graphs (DAGs), where each DAG contains all candidate operations. First, the Supernet is trained and updated by clients with fixed architecture parameters on mini-batches of data. Then, the architecture parameters are updated with fixed model parameters on mini-batch validation data. Clients will alternate between training and updating the architecture during one local training epoch. All participating clients upload both model and architecture parameters to the server after training for several epochs. The server then averages the received Supernets to create a new global one. While this method is less communication heavy, it requires had higher computation and memory requirements, which is not suitable for many resource-constrained devices. Several enhancements were therefore proposed such as using model sampling [86] and meta-learning [88]. Notably, in model sampling, only one path of cells in the Supernet is sampled and sent to local clients. In meta-learning, the weights of a pretrained DNN can have a high transferability when used in new tasks, allowing each device to learn a model more suitable for its particular hardware and data distribution.

3.5 GREEN FEDERATED LEARNING PROCESS

In addition to the careful model design, the different steps of the learning process should be tailored in an energy-efficient manner, from client selection and local training to model upload, aggregation, and personalization. Figure 3.4 provides a roadmap for the different techniques and optimization methods that can be leveraged for green FL, which we discuss in detail in this section.

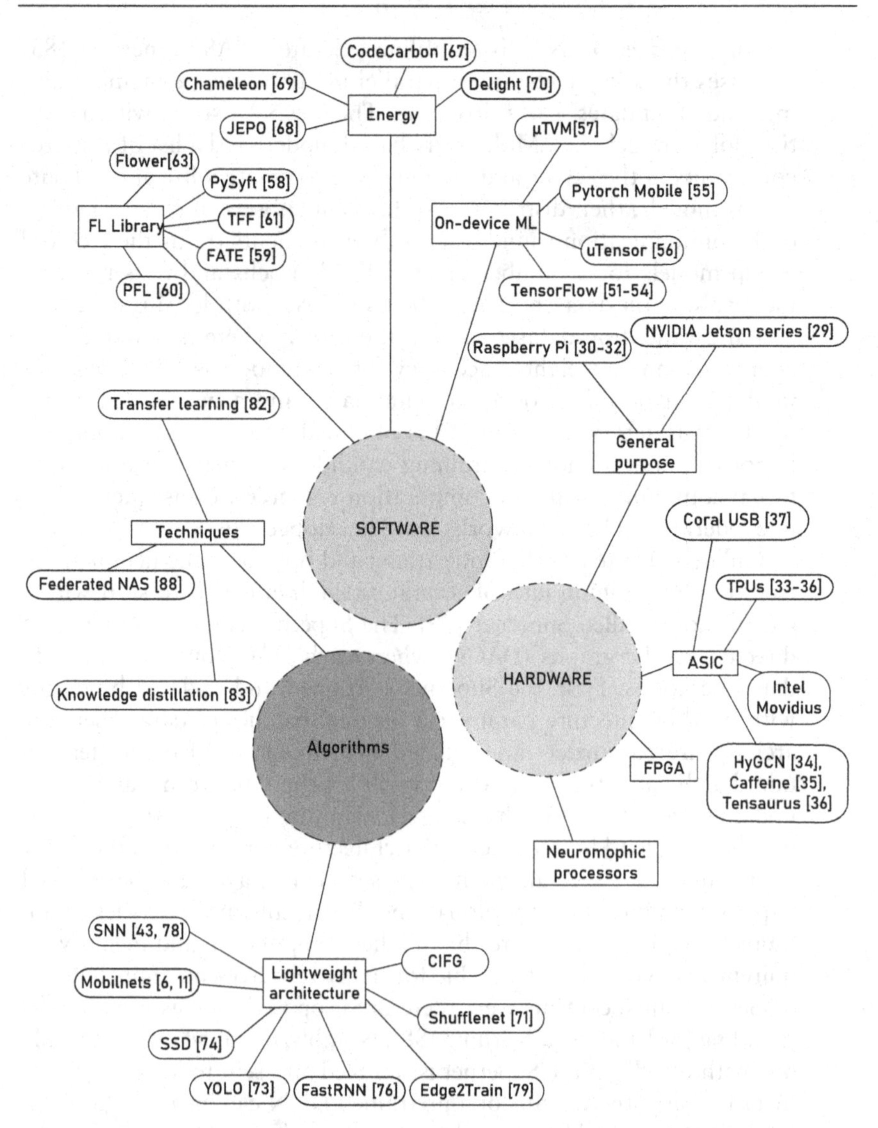

Figure 3.4 Green FL process optimization techniques in the different steps of the iterative process to slow down after a few iterations [90].

3.5.1 Client selection

Ideally in FL, all the clients participate in the training process. However, due to resource constraints, only a subset can participate. From an energy perspective, only devices plugged into an energy source or with enough battery for training and upload should participate for example [14, 15]. The client selection step is often coupled with resource allocation for more efficient usage of the resources, through optimal selection of training (e.g., CPU frequency) and upload parameters (e.g., transmission power). Additionally,

Table 3.1 Summary of client selection strategies

Strategy	*Main goal*	*Limitations*
Maximizing the spectral utility [91, 92]	Maximize the number of clients participating in each round	Biased toward powerful devices
Round Robin / Proportional Fair [18], and age-based scheduling [93]	Diversify the sources of the updates and introduce fairness in the selection	Low participation and device-drop
Active-learning based methods[10, 94]	Select clients with potentially more informative datasets and training samples	Low participation
Constant number of participants [95]	Maintain stable training	Not always possible due to varying numbers of available clients
Selection with long-term perspective [90]	Schedule more clients in later rounds to save energy	Later rounds might take longer and require more communication resources

energy harvesting devices can be leveraged for sustainable training either as alternative energy sources to avoid device dropout due to battery drainage, or as the main clients. However, energy harvesting devices require new considerations due to the stochastic energy arrival aspects. Table 3.1 summarizes different client selection strategies and their limitations.

3.5.1.1 Client selection and resource allocation

Bandwidth allocation can consist of allocation of the upload time [92] in the case of Time Division Multiple Access (TDMA), fraction of the bandwidth [16] or number of resource blocks [96] in the case of Orthogonal frequency-division multiple access (OFDMA), and/or transmission power [17] for non-orthogonal multiple access (NOMA). The client selection and resource allocation problem can, for instance, be formulated with the aim to optimize energy consumption across devices [97], where adequate transmission power and CPU-frequency are chosen.

The convergence analysis of FL suggests that maintaining a balanced number of participants in each round is desirable [95]. Some works set maximizing the number of participants in each round as the goal of their scheduling algorithms [97]. However, in the case of battery-powered devices where the energy budget is limited, more participants should be included in later rounds rather than earlier rounds, as the training tends

Nonetheless, a clear trade-off can between time and energy is observed. Specifically, in order to train and upload models in short amounts of time, devices will need to use the highest CPU frequency and high transmission power, which entails high energy consumption. Many proposed algorithms consider both time and energy during resource allocation [98]. For instance, the resource allocation can be formulated as a multi-objective optimization problem, aiming to minimize both time and energy within a resource budget [99]. In general, defining the client selection and resource allocation problem will heavily depend on the use-case at hand and the environment's constraints. In the case of battery-powered devices for instance, we can define an optimization problem with energy-related constraints [100, 101], while for devices that are always plugged in, the focus might be on minimizing the round duration.

3.5.1.2 Client selection and energy harvesting devices

Since battery-powered devices are prone to drop-out, and diversifying the updates' sources is important for FL performance over non-i.i.d. data, energy harvesting technology is used prevent device drop-outs by providing an extra-source of energy [102, 103]. Nonetheless, energy harvesting devices are also subject to uncertainty of energy arrivals. To overcome challenges posed by the intermittent energy availability for energy harvesting devices, a potential approach is to select clients for participation as soon as they generate enough energy. However, this approach might be biased toward clients with faster energy generation and result in global models that do apply poorly on the remainder of clients. Another possible solution would be to wait until all clients generate sufficient energy to carry the training and upload tasks. While this might reduce the training bias, it will entail significant delays as it will require waiting for the clients with the slowest energy arrival rate [104]. A tractable approach could be to maintain a constant number of participants in each round. Nonetheless, stochastic energy arrival at client devices makes maintaining a constant number a challenging task [95]. The client selection in this case can prioritize clients with the longest energy queues and the long-term average energy harvesting rate [105].

3.5.1.3 Client selection with non-i.i.d. data

While the above presented client selection approaches focus on optimizing the resources, the statistical heterogeneity of the data is often overlooked. As a result, in non-i.i.d. distributions, the FL's convergence is significantly slowed as the client selection is biased toward a subset of devices with more powerful processing units and higher transmission power. Instead, data and model properties should be considered during client selection [106]. This could be achieved through defining a combined statistical and system utility [94] or considering the client selection as a multi-objective goal aiming to reduce energy and maximizing the data diversity [10]. Such careful client selection can also be reinforced with optimized local training through a judicious training samples selection [107].

3.5.2 Local training

In many cases, FL clients' energy consumption is dominated by ML training which can be several times more energy-hungry than model upload, irrespective of the communication channel. Large local datasets can deplete a high percentage of device's battery, thus discouraging client participation in FL [108]. It is therefore necessary to optimize the training process through a careful choice of training parameters and developing partial training algorithms. Table 3.2 summarizes the different approaches used to optimize the local training.

3.5.2.1 Training parameters

The main idea behind FL is to trade communication with more local computation in order to avoid the communication bottleneck. However, training itself is a computation heavy task. As a result, optimizing training (e.g., number of local iterations, batch size) and uploading parameters is necessary for the energy efficiency of the communication round. For instance, [22] the numbers of global and local iterations can be chosen depending on the available communication and computation resources as a means to better profit from the trade-off. Other parameters such as batch size enforced by appropriate learning rates can be adaptively chosen for each device [109]. The energy consumption of local training and wireless communication from the long-term learning perspective can also be enhanced through the usage of compression techniques. In particular, the compression parameters need to be judiciously selected for FL participants depending on their computing resources and communication environments [110].

Table 3.2 Summary of local training optimization techniques

Technique	*Main goal*	*Limitations*
Optimizing training parameters	Select batch-size [109], number of epochs and number of communication rounds [110]	Requires new aggregation schemes
Sketched updates [7]	Only update a subset of parameters locally	Slows the training convergence
Transfer learning [82]	Train the last layers locally	Requires the availability of a similar model
Model Distillation [83]	Different model for each client to handle resourcesheterogeneity	Requires a shared dataset

For energy harvesting devices, intermittent learning is introduced to adapt to the heterogeneous energy arrival rates [111]. In this setting, a learning task is divided into several possible actions, namely sensing, selecting, learning, and inferring. The heuristic-based algorithm selects the action that maximizes learning performance with respect to energy constraints. The proposed system also decides which training samples to use for training based on their potential to improve the learning performance, all within the energy budget.

3.5.2.2 Partial training

In addition to the careful choice of the hyperparameters, the training process can be enabled through partial training. The partial training can either mean training only a subset of parameters or training on a subset of the data.

Training a subset of the parameters can be achieved through structured updates [7], which consist of restricting the model update to be a sparse matrix following a predefined random sparsity pattern where a subset of neurons in an ANN are masked. The pattern is generated afresh by a random seed in each round and for each client independently. The sparsity pattern can also be dynamically chosen for each client so as to take resources heterogeneity and limitations into account, where the subsets of neurons of an ANN are to be masked in each round for each client based on energy and time profiling [112].

Training on a subset of the samples is mainly a means to avoid redundant or mislabeled samples. Indeed, redundancy of local datasets entails unnecessary training on samples that do not improve the training. Consequently, several methods based on Active learning can be used to select the samples that enhance the model's quality based [107]. Furthermore, another factor that can incur additional energy costs is missing or wrong labels in training data. The common underlying assumption of current FL applications is that all data samples are appropriately labeled. However, in reality, data collected by end devices is highly subject to imperfections and malicious entities. To avoid squandering energy on such samples, different mitigation methods can be used. Nonetheless, these measures can also entail additional energy costs.

3.5.3 Updates collection

As communication is often the bottleneck of FL, updates collection is often considered a resource consuming step. As a result, various solutions were studied and developed. Particularly, several preexisting compression methods, alongside new partial updates collection techniques, were proposed. Table 3.3 summarizes the different approaches for communication efficiency in the updates collection step.

Table 3.3 Summary of model upload optimization techniques

Technique	*Main goal*	*Limitations*
Quantization [12, 113]	Changing the model's parameters from floating-point numbers to low-bit width numbers	Accuracy drop
Pruning [12]	Removing the least important parameters	Accuracy drop
Sketched updates [7]	Send a subset of parameters only	Longer time to convergence
Federated distillation [114]	upload per-label mean logit vectors	Privacy concerns

3.5.3.1 Compression techniques

Many compression techniques that enabled inference and communication efficiency in ML can be also exploited for green FL such as quantization and pruning.

Quantization [13] consists of changing the model's parameters from floating-point numbers to low-bit width numbers thus reducing the cost of multiplication operations. For instance, we can reduce a model's size by a factor of 4, by approximating 32-bit floating-point numbers with 8-bit numbers. In order to adequately implement quantization in FL, different quantization levels can be used throughout the communication rounds [115]. A suitable choice of the quantization level is necessary to avoid high error rates due to the higher variance of the gradients. Furthermore, coupled with an appropriate aggregation method, different quantization levels can be used across clients to adjust with the devices' heterogeneity [113].

Pruning [12] is also a widely adopted technique of model compression, used in neural networks and decision-trees. Pruning a neural network consists of removing the least important parameters (e.g., weights below a threshold). Nonetheless, removing neurons might damage the accuracy of the DNN, leading to an important trade-off between the network size and accuracy. Moreover, pruning may not always be beneficial, since empirical measurements show that the reduction of the number of weights does not necessarily help energy saving. For instance, in CNNs, the energy consumption of the convolutional have the highest energy cost, even if the fully connected layers constitute the majority of the neurons and connections. Pruning can be used in FL [116] for both training and upload. However, a pruning policy for a given model depends highly on the local dataset, which is not suitable for non-i.i.d. settings, thus requiring new considerations and techniques [117].

3.5.3.2 Partial updates

Similarly to partial training, partial updates can be used to reduce the energy consumption during the model upload, particularly in cases where the training is much less resource-hungry compared to upload (e.g., very few local samples). For instance, shallow layers learn general features in contrast to deep layers that learn more specific features. As a result, shallow layers can be aggregated more frequently to build a global model in a more energy-efficient manner [118]. Another approach is sketched updates [7], which, similar to structured updates, consist of sending only a subset of parameters using a random mask. Combined with compression techniques such as quantization and pruning, structured updates achieve high savings in communication cost.

Due to the important accuracy-energy efficiency trade-off present with gradient compression, the compression ratio and computing frequency of each device can be jointly optimized. To adapt to the heterogeneous resources, each device can adopt a different compression ratio on demand alongside an adequate computing frequency [119].

3.5.3.3 Federated distillation

An interesting communication-efficient solution is federated distillation (FD) [114], as it proposes an alternative to sending entire model updates inspired from knowledge distillation. In FD, each client stores per-label mean logit vectors (i.e., the average of raw and non-normalized predictions that the local model generates), and periodically uploads them to a server instead of model updates. The mean logit vectors from all clients are averaged for each label and the global-average logit vectors of all labels are sent back to the clients. Then, each client calculates the distillation regularizer, where the teacher's output is selected as the global-average logit vector corresponding to training sample's label. By sharing the model outputs instead of the entire gradient, the communication overhead is significantly reduced.

3.5.4 Models aggregation

How often the models are aggregated and how they are aggregated substantially affect the energy efficiency. In particular, using an intermediate aggregation layer using edge equipment can reduce the energy consumption compared to directly sending updates to the cloud. Additionally, adopting asynchronous aggregation and reducing the updates collections frequency are also considered as manageable solutions adding more flexibility to the FL process, especially the updates collection step where communication is considered the bottleneck of FL.

3.5.4.1 Hierarchical aggregation

Introducing an intermediate aggregation operation at the edge server level before the global aggregation at the central server can lower the communication cost. This can be achieved by leveraging the preexisting client-edge-cloud network topology [120]. As a result, better communication-computation trade-off can be achieved as the communication efficiency is significantly enhanced. Particularly, it is shown that the client-edge-cloud training reduces the model training time and the energy consumption of the end devices compared to directly sending updates to cloud servers [121]. Energy efficiency can further be improved in hierarchical FL through communication and computation optimization in each tier [122].

3.5.4.2 Asynchronous aggregation

To overcome issues related to clients' availability (e.g., energy harvesting devices), and other challenges that stem from resource heterogeneity such as the straggler's problem, a popular solution is asynchronous aggregation. In this setting, the server performs aggregation upon receiving any update, and may include a new client whenever it is available. Collecting the updates in an asynchronous manner and assigning weights depending on the staleness of the update allow a smooth adaptation to heterogeneous resources through a flexible updates collection [123]. The weight assigned to the updated gradients can also be set to decrease as the staleness value increases [124]. Asynchronous aggregation can also be coupled with caching so as to maximize the number of collected updates. For instance, if a client is not able to upload its local model due to lack of energy, the update can be cached locally for several iterations [125]. Nonetheless, the stale updates can significantly harm the training. Furthermore, the convergence is theoretically guaranteed only on a subset of problems [123]. As a result, synchronous aggregation remains the most adopted method [2].

3.5.4.3 Important and diverse updates

Due to redundancy in user-generated data, various works consider that not all updates are important. For instance, Gaia [126] determines which local updates should not be uploaded based on their magnitude compared to the global model. However, this method cannot be applied for non-i.i.d. distributions and under the possibility of poisoning attacks. A significance measure based on the percentage of same-sign parameters in the two updates was proposed [19]. This approach identifies whether the client-side updates follow the more general and common direction and excludes outliers. Evaluating the updates' significance might help mitigate the communication overhead and improve the learning efficiency by excluding devices with redundant data and outliers. However, this method is not suitable in energy and resource-constrained environments, mainly leading to wasting devices computing resources and energy. Instead, careful client selection prior to training and active-learning-based approaches based on careful construction of local datasets are to be encouraged for more sustainable FL.

3.5.5 Personalization

As the statistical heterogeneity challenges are often exacerbated due to other challenges such as concept-shift, a single global model will not perform well for all participants. As a result, a number of personalization techniques were proposed. The first proposed approach was to retrain the models locally for each client. Nonetheless, it becomes quickly clear that this method might lead to overfitted models together with additional computation costs. Transfer learning can be executed locally by freezing the base layers of the global model and retraining only the top layers on local data [127]. It was also proposed to consider the global model as the teacher and the personalized local model as the student, the effects of overfitting during personalization can be mitigated [83]. This approach can lead to lighter models in a more sustainable manner.

Another approach is to add user context to obtain personalized models. This can be simply achieved through clustering clients by grouping them using some general features such as the geographical location. Such grouping can leverage edge equipment to bypass the communication bottleneck. Multitask learning was also shown to be a natural choice for personalization is FL, as it aims to learn multiple tasks simultaneously allowing the global model to learn known and hidden commonalities between tasks and train on them simultaneously. MOCHA [128] was developed to tackle challenges related to communication, stragglers, and device dropout in multitask FL. However, their evaluations show that partial participation significantly harms the training. Meta-learning is also an interesting solution for personalization [129]. The idea behind meta-learning is to train models on multiple learning tasks to generate highly adaptable models that can later be reused on new tasks where there is only a small number of training samples. The similarity between meta-learning and FL becomes clear if we consider the FL process as meta-training, as the personalization process can be seen as equivalent to meta-testing. Accordingly, previously used methods to optimize meta-learning and meta-testing can be exploited for energy-efficient personalization.

3.6 OPEN CHALLENGES AND FUTURE DIRECTIONS

Despite existing efforts toward resource efficiency in FL, the road toward green FL is still long and presents many challenges ahead. In particular, an end-to-end green FL process will need to account for unlabeled and untrustworthy data and develop privacy-efficient algorithms.

3.6.1 Unlabeled data

Most of the discussed techniques in this chapter focus on supervised learning. The premise of supervised learning is that all data is already labeled. In practice, the data generated in the network might be unlabeled or mislabeled. Consequently, it is necessary to rethink FL algorithms to take into account unlabeled data and to include semi-supervised and unsupervised learning algorithms. However, this might require new methods to adapt to system heterogeneity, scarcity, and privacy challenges.

3.6.2 Trust and auditability

FL is vulnerable to client dropout, noisy channels, poisoning attacks, and other misbehaviors. Such incidents and attacks are not auditable, and the global model might be incorrectly computed due to single point failure. Several blockchain-based processes were proposed to enable trust among the different participants [130–133]. In such settings, local updates are exchanged and verified and utilizes a consensus mechanism in blockchain. However, these approaches require that the clients be also miners. As the training and upload are already energy-hungry tasks, cross-verification and mining incur additional costs and operations that go against making FL sustainable. As a result, lightweight auditability mechanisms are necessary to protect FL and keep the process green.

3.6.3 Energy-efficient privacy algorithms

Existing approaches applied to FL to further promote data privacy consist of adapting classical cryptographic protocols and algorithms, namely secure multiparty computation, and differential privacy. However, secure multiparty computation entails high communication costs which do not align with green FL goals. A promising alternative is secure aggregation [134], where a virtual and trusted third-party server is deployed for model aggregation, as runtime and communication overhead remain low even on large data sets and client pool. Another popular direction is differential privacy. Differential privacy is based on adding a certain degree of noise (e.g., Gaussian noise) to the local updates while providing theoretical guarantees on the model quality. Yet, in practice, the adoption of these approaches sacrifices the performance and requires significant computation mobile devices. Thus, the trade-off between privacy guarantee and computation cost has to be balanced when developing FL protocols.

3.6.4 End-to-end energy-efficient FL protocol

While a large landscape of studies about communication and computation efficiency has been proposed and studied, many focus on only a couple of steps in FL at best. The feasibility and efficiency of a green FL protocol through a combination of these techniques and their impact on the accuracy needs to be further evaluated. In particular, a combination of careful model design, compression techniques, and privacy enhancing techniques could significantly make the model lighter and achieve FL's privacy goals; however, the convergence guarantees, and the model accuracy need to be studied under such combined protocol.

3.7 CONCLUSION

FL enables training shared ML models across distributed client devices while keeping all the training data on-device. This extends the edge intelligence capabilities beyond the use of models for inference to on-device training.

Recently, the topic of FL has gained a growing interest from both industry and academia owing to its privacy-promoting aspect. While few companies have already successfully deployed FL in production, a wider adoption is conditional to the optimization of FL to resource constrained devices. In particular, training models is a source of important CO2 emissions, and the development of sustainable FL is of upmost importance. Nonetheless, several challenges need to be addressed in relation to system and data heterogeneity and scarcity. To this end, different existing technologies in terms of software and hardware can be leveraged and others need to be further developed. For instance, FL libraries need to be extended with lightweight privacy and security functionalities. Furthermore, lightweight model architectures and algorithms can be used to obtain models with reduced energy requirements. Notably, modifications to neural networks architectures, such as simplified convolutions, and parallelized operations in CNNs and recurrent units with fewer parameters. Moreover, other techniques like transfer learning and knowledge distillation can be key techniques for model initialization, and online federated network architecture search can be used to find suitable model architecture under heterogeneous settings.

Optimizing the global model is not enough to reduce the energy required for FL, as model design is only the first step in the process. Instead, each different step of FL should be crafted with energy efficiency in mind. Client selection in particular has been the focus of several research efforts. This step is often coupled with resource allocation, and requires supplementary measures for energy harvesting devices and non-i.i.d. data, so as to adapt to the low availability of the clients and their heterogeneous data. The next step to be optimized is the local training, where the training parameters and enabling partial training are vital to decrease the energy requirements. For the updates collection step, many compression techniques such as quantization can be adapted to device heterogeneity to reduce the communication cost, and new partial updates methods like sketched updates can be used to reduce the communication overhead. Additionally, how and where the models are aggregated play a significant part in a green FL process. For instance, introducing an intermediate aggregation operation in edge servers and enabling asynchronous FL can substantially enhance energy efficiency in FL. Lastly, personalization techniques also incur more energy costs but are essential to enhance local accuracy for heterogeneous clients. Personalization can be designed in an energy-efficient fashion by exploiting existing ideas such as transfer learning and meta-learning.

Despite the existing efforts toward sustainable and green FL, many aspects of FL require new considerations. Notably, green FL protocols that handle auditability, enhanced privacy, unlabeled and missing data need to be developed.

NOTE

1 https://codecarbon.io/

REFERENCES

[1] H. B. McMahan, E. Moore, D. Ramage, S. Hampson, and B. A. Y. Arcas, "Communication-Efficient Learning of Deep Networks from Decentralized Data," *arXiv:1602.05629 [cs]*, Feb. 2017. [Online]. Available: http://arxiv.org/abs/1602.05629

[2] K. Bonawitz et al., "Towards Federated Learning at Scale: System Design," 2019.

[3] C. Wang, X. Wei, and P. Zhou, "Optimize Scheduling of Federated Learning on Battery-Powered Mobile Devices," in *2020 IEEE International Parallel and Distributed Processing Symposium (IPDPS)*, IEEE, 2020, pp. 212–221.

[4] J. Chen, X. Pan, R. Monga, S. Bengio, and R. Jozefowicz, "Revisiting Distributed Synchronous SGD," *arXiv preprint arXiv:1604.00981*, 2016.

[5] A. Hard, K. Rao, R. Mathews, S. Ramaswamy, F. Beaufays, S. Augenstein, H. Eichner, C. Kiddon, and D. Ramage, "Federated Learning for Mobile Keyboard Prediction," *arXiv preprint arXiv:1811.03604*, 2018.

[6] M. Sandler, A. Howard, M. Zhu, A. Zhmoginov, and L.-C. Chen, "Mobilenetv2: Inverted Residuals and Linear Bottlenecks," in *Proceedings of the IEEE conference on computer vision and pattern recognition*, 2018, pp. 4510–4520.

[7] J. Konecnˇ, Y. H. B. McMahan, F. X. Yu, P. Richtárik, A. T. Suresh, and D. Bacon, "Federated Learning: Strategies for Improving Communication Efficiency," *arXiv:1610.05492 [cs]*, Oct. 2017. [Online]. Available: http://arxiv.org/abs/1610.05492

[8] N. Zmora, G. Jacob, L. Zlotnik, B. Elharar, and G. Novik, "Neural Network Distiller: A Python Package for DNN Compression Research," *arXiv preprint arXiv:1910.12232*, 2019.

[9] Y. Zhao, M. Li, L. Lai, N. Suda, D. Civin, and V. Chandra, "Federated Learning with Non-IID Data," *arXiv preprint arXiv:1806.00582*, 2018.

[10] A. Taïk, Z. Mlika, and S. Cherkaoui, "Data-aware Device Scheduling for Federated Edge Learning," *IEEE Transactions on Cognitive Communications and Networking*, pp. 1–1, 2021.
[11] A. G. Howard, M. Zhu, B. Chen, D. Kalenichenko, W. Wang, T. Weyand, M. Andreetto, and H. Adam, "MobileNets: Efficient Convolutional Neural Networks for Mobile Vision Applications," *arXiv:1704.04861 [cs]*, Apr. 2017, arXiv: 1704.04861. [Online]. Available: http://arxiv.org/abs/1704.04861
[12] W. Xu, W. Fang, Y. Ding, M. Zou, and N. Xiong, "Accelerating Federated Learning for IoT in Big Data Analytics With Pruning, Quantization and Selective Updating," *IEEE Access*, vol. 9, pp. 38457–38466, 2021, conference Name: IEEE Access.
[13] N. Shlezinger, M. Chen, Y. C. Eldar, H. V. Poor, and S. Cui, "Federated Learning with Quantization Constraints," in *ICASSP 2020 - 2020 IEEE International Conference on Acoustics, Speech and Signal Processing (ICASSP)*, 2020, pp. 8851–8855.
[14] Y. Sun, S. Zhou, and D. Gündüz, "Energy-Aware Analog Aggregation for Federated Learning with Redundant Data," in *ICC 2020 - 2020 IEEE International Conference on Communications (ICC)*, Jun. 2020, pp. 1–7, iSSN: 1938-1883.
[15] C. Wang, X. Wei, and P. Zhou, "Optimize Scheduling of Federated Learning on Battery-powered Mobile Devices," in *2020 IEEE International Parallel and Distributed Processing Symposium (IPDPS)*, May 2020, pp. 212–221, iSSN: 1530-2075.
[16] Q. Zeng, Y. Du, K. K. Leung, and K. Huang, "Energy-Efficient Radio Resource Allocation for Federated Edge Learning," *arXiv:1907.06040 [cs, math]*, Jul. 2019. [Online]. Available: http://arxiv.org/abs/1907.06040
[17] B. Li, A. Swami, and S. Segarra, "Power Allocation for Wireless Federated Learning Using Graph Neural Networks," 2021.
[18] H. H. Yang, Z. Liu, T. Q. S. Quek, and H. V. Poor, "Scheduling Policies for Federated Learning in Wireless Networks," *IEEE Transactions on Communications*, vol. 68, no. 1, pp. 317–333, 2020.
[19] L. Wang, W. Wang, and B. Li, "CMFL: Mitigating Communication Overhead for Federated Learning," in *IEEE ICDCS 2019*, Dallas, TX, USA, Jul. 2019, pp. 954–964. [Online]. Available: https://ieeexplore.ieee.org/document/8885054/
[20] C. Xu, Y. Qu, Y. Xiang, and L. Gao, "Asynchronous Federated Learning on Heterogeneous Devices: A Survey," *arXiv preprint arXiv:2109.04269*, 2021.
[21] P. Kairouz et al., "Advances and Open Problems in Federated Learning," 2021.
[22] S. Wang, T. Tuor, T. Salonidis, K. K. Leung, C. Makaya, T. He, and K. Chan, "Adaptive Federated Learning in Resource Constrained Edge Computing Systems," *IEEE Journal on Selected Areas in Communications*, vol. 37, no. 6, pp. 1205–1221, Jun. 2019.
[23] T. Li, A. K. Sahu, A. Talwalkar, and V. Smith, "Federated Learning: Challenges, Methods, and Future Directions," *IEEE Signal Processing Magazine*, vol. 37, no. 3, pp. 50–60, 2020.
[24] A. Reisizadeh, I. Tziotis, H. Hassani, A. Mokhtari, and R. Pedarsani, "Stragglerresilient Federated Learning: Leveraging the Interplay between Statistical Accuracy and System Heterogeneity," *arXiv preprint arXiv:2012.14453*, 2020.

[25] R. Hamdi, M. Chen, A. B. Said, M. Qaraqe, and H. V. Poor, "User Scheduling in Federated Learning Over Energy Harvesting Wireless Networks," in *2021 IEEE Global Communications Conference (GLOBECOM)*. IEEE, 2021, pp. 1–6.
[26] H. Shen, M. Philipose, S. Agarwal, and A. Wolman, "MCDNN: An Execution Framework for Deep Neural Networks on Resource-Constrained Devices," *Proceedings of the 14th Annual International Conference on Mobile Systems, Applications, and Services*, pp. 123–136, 2016.
[27] E. Ragusa, C. Gianoglio, R. Zunino, and P. Gastaldo, "A Design Strategy for the Efficient Implementation of Random Basis Neural Networks on Resource Constrained Devices," *Neural Processing Letters*, vol. 51, no. 2, pp. 1611–1629, Apr. 2020. [Online]. Available: https://doi.org/10.1007/s11063-019-10165-y
[28] J. Jiang and L. Hu, "Decentralised federated learning with adaptive partial gradient aggregation," *CAAI Transactions on Intelligence Technology*, vol. 5, no. 3, pp. 230–236, May 2020.
[29] S. Alyamkin, M. Ardi, A. C. Berg, A. Brighton et al., "Low-Power Computer Vision: Status, Challenges, and Opportunities," *IEEE Journal on Emerging and Selected Topics in Circuits and Systems*, vol. 9, no. 2, pp. 411–421, Jun. 2019, conference Name: IEEE Journal on Emerging and Selected Topics in Circuits and Systems.
[30] S. Y. Nikouei, Y. Chen, S. Song, R. Xu, B.-Y. Choi, and T. Faughnan, "Smart Surveillance as an Edge Network Service: From Harr-cascade, SVM to a Lightweight CNN," in *2018 IEEE 4th International Conference on Collaboration and Internet Computing (CIC)*, 2018, pp. 256–265.
[31] R. Xu, S. Y. Nikouei, Y. Chen, A. Polunchenko, S. Song, C. Deng, and T. R. Faughnan, "Real-time Human Objects Tracking for Smart Surveillance at the Edge," in *2018 IEEE International Conference on Communications (ICC)*, 2018, pp. 1–6.
[32] N. Besimi, B. Çiço, A. Besimi, and V. Shehu, "Using Distributed Raspberry PIs to Enable Low-Cost Energy-Efficient Machine Learning Algorithms for Scientific Articles Recommendation," *Microprocessors and Microsystems*, vol. 78, p. 103252, Oct. 2020. [Online]. Available: https://www.sciencedirect.com/science/article/pii/S0141933120304130
[33] E. Shapiro, "Systolic Programming: A Paradigm of Parallel Processing," in *Concurrent Prolog*, 1988, pp. 207–242.
[34] M. Yan, L. Deng, X. Hu, L. Liang, Y. Feng, X. Ye, Z. Zhang, D. Fan, and Y. Xie, "Hygcn: A GCN Accelerator with Hybrid Architecture," in *2020 IEEE International Symposium on High Performance Computer Architecture (HPCA)*, 2020, pp. 15–29.
[35] C. Zhang, G. Sun, Z. Fang, P. Zhou, P. Pan, and J. Cong, "Caffeine: Toward Uniformed Representation and Acceleration for Deep Convolutional Neural Networks," *IEEE Transactions on Computer-Aided Design of Integrated Circuits and Systems*, vol. 38, no. 11, pp. 2072–2085, 2018.
[36] N. Srivastava, H. Jin, S. Smith, H. Rong, D. Albonesi, and Z. Zhang, "Tensaurus: A Versatile Accelerator for Mixed Sparse-Dense Tensor Computations," in *2020 IEEE International Symposium on High Performance Computer Architecture (HPCA)*. IEEE, 2020, pp. 689–702.
[37] "AlasdairAllan," https://aallan.medium.com/benchmarking-edge-computing-ce3f13942245, accessed: 2022-02-27.

[38] J. Jiang, C. Liu, and S. Ling, "An FPGA Implementation for Real-Time Edge Detection," *Journal of Real-Time Image Processing*, vol. 15, no. 4, pp. 787–797, 2018.

[39] G. Devic, G. Sassatelli, and A. Gamatie, "Energy-Efficient Machine Learning on FPGA for Edge Devices: A Case Study," in *ComPAS: Conference en Parall elisme, Architecture et Systeme*, 2020.

[40] C. Lammie, A. Olsen, T. Carrick, and M. R. Azghadi, "Low-Power and Highspeed Deep FPGA Inference Engines for Weed Classification at the Edge." *IEEE Access*, vol. 7, pp. 51171–51184, 2019.

[41] L. Lai and N. Suda, "Enabling Deep Learning at the IOT Edge," in *Proceedings of the International Conference on Computer-Aided Design, ser. ICCAD '18*. New York, NY, USA: Association for Computing Machinery, 2018. [Online]. Available: https://doi.org/10.1145/3240765.3243473

[42] S. Furber, "Large-scale Neuromorphic Computing Systems," *Journal of Neural Engineering*, vol. 13, no. 5, p. 051001, Aug. 2016, publisher: IOP Publishing. [Online]. Available: https://doi.org/10.1088/1741-2560/13/5/051001

[43] K. Stewart and Y. Gu, "One-Shot Federated Learning with Neuromorphic Processors," *arXiv:2011.01813 [cs]*, Nov. 2020, arXiv: 2011.01813. [Online]. Available: http://arxiv.org/abs/2011.01813

[44] M. Bouvier, A. Valentian, T. Mesquida, F. Rummens, M. Reyboz, E. Vianello, and E. Beigne, "Spiking Neural Networks Hardware Implementations and Challenges: A Survey," *ACM Journal on Emerging Technologies in Computing Systems*, vol. 15, no. 2, pp. 1–35, Jun. 2019. [Online]. Available: https://dl.acm.org/doi/10.1145/3304103

[45] M. Davies, N. Srinivasa, T.-H. Lin et al., "Loihi: A Neuromorphic Manycore Processor with On-chip Learning," *IEEE Micro*, vol. 38, no. 1, pp. 82–99, 2018.

[46] P. A. Merolla, J. V. Arthur, R. Alvarez-Icaza, A. S. Cassidy, J. Sawada, F. Akopyan, B. L. Jackson, N. Imam, C. Guo, Y. Nakamura, B. Brezzo, I. Vo, S. K. Esser, R. Appuswamy, B. Taba, A. Amir, M. D. Flickner, W. P. Risk, R. Manohar, and D. S. Modha, "A Million Spikingneuron Integrated Circuit with a Scalable Communication Network and Interface," *Science*, vol. 345, no. 6197, pp. 668–673, Aug. 2014, publisher: American Association for the Advancement of Science. [Online]. Available: https://www.science.org/doi/full/10.1126/science.1254642

[47] P. Levis, S. Madden, J. Polastre, R. Szewczyk, K. Whitehouse, A. Woo, D. Gay, J. Hill, M. Welsh, E. Brewer, and D. Culler, "TinyOS: An Operating System for Sensor Networks," in *Ambient Intelligence*, W. Weber, J. M. Rabaey, and E. Aarts, Eds. Berlin, Heidelberg: Springer, 2005, pp. 115–148.

[48] Z. Xu, X. Peng, L. Zhang, D. Li, and N. Sun, "The Phi-Stack for Smart Web of Things," in *Proceedings of the Workshop on Smart Internet of Things, ser. SmartIoT 17*. New York, NY, USA: Association for Computing Machinery, Oct. 2017, pp. 1–6. [Online]. Available: https://doi.org/10.1145/3132479.3132489

[49] E. García-Martín, C. F. Rodrigues, G. Riley, and H. Grahn, "Estimation of Energy Consumption in Machine Learning," *Journal of Parallel and Distributed Computing*, vol. 134, pp. 75–88, Dec. 2019. [Online]. Available: https://www.sciencedirect.com/science/article/pii/S0743731518308773

[50] X. Zhang, Y. Wang, and W. Shi, "{pCAMP}: Performance Comparison of Machine Learning Packages on the Edges," 2018. [Online]. Available: https://www.usenix.org/conference/hotedge18/presentation/zhang

[51] "Keras: The Python deep learning API." [Online]. Available: https://keras.io/
[52] "TensorFlow I/O." [Online]. Available: https://www.tensorflow.org/io
[53] "TensorFlow Lite | ML for Mobile and Edge Devices." [Online]. Available: https://www.tensorflow.org/lite
[54] "TensorFlow Privacy | Responsible AI Toolkit." [Online]. Available: https://www.tensorflow.org/responsibleai/privacy/guide
[55] "PyTorch Mobile." [Online]. Available: https://www.pytorch.org/mobile/home/
[56] "uTensor - Test Release," Jul. 2022, original-date: 2017-09-21T17:04:59Z. [Online]. Available: https://github.com/uTensor/uTensor
[57] "microTVM: TVM on bare-metal — tvm 0.9.dev0 documentation." [Online]. Available: https://tvm.apache.org/docs/topic/microtvm/index.html
[58] A. Ziller, A. Trask, A. Lopardo, B. Szymkow, B. Wagner, E. Bluemke, J.-M. Nounahon, J. Passerat-Palmbach, K. Prakash, N. Rose et al., "Pysyft: A Library for Easy Federated Learning," in *Federated Learning Systems*. Springer, 2021, pp. 111–139.
[59] FedAI, "FedAI Home." [Online]. Available: https://www.fedai.org/
[60] "PaddlePaddle/PaddleFL," Jul. 2022, original-date: 2019-09-25T15:01:39Z. [Online]. Available: https://github.com/PaddlePaddle/PaddleFL
[61] "TensorFlow Federated." [Online]. n.d. Available: https://www.tensorflow.org/federated
[62] K. Bonawitz, H. Eichner, W. Grieskamp et al., "Tensorflow Federated: Machine Learning On Decentralized Data. (2020)."
[63] T. F. Authors, n.d. "Flower: A Friendly Federated Learning Framework." [Online]. Available: https://flower.dev/
[64] R. A. Bridges, N. Imam, and T. M. Mintz, "Understanding GPU Power: A Survey of Profiling, Modeling, and Simulation Methods," *ACM Computing Surveys (CSUR)*, vol. 49, no. 3, pp. 1–27, 2016.
[65] C. F. Rodrigues, G. Riley, and M. Lujan, "SyNERGY: An Energy Measurement and Prediction Framework for Convolutional Neural Networks on Jetson TX1," *Undefined*, 2018. [Online]. Available: https://www.semanticscholar.org/paper/SyNERGY%3A-An-energy-measurement-and-prediction-for-Rodrigues-Riley/cca7a56ae73ab0428e9859146f222a750a390c7d
[66] R. W. Ahmad, A. Gani, S. H. A. Hamid, F. Xia, and M. Shiraz, "A Review on Mobile Application Energy Profiling: Taxonomy, State-of the-Art, and Open Research Issues," *Journal of Network and Computer Applications*, vol. 58, pp. 42–59, 2015. [Online]. Available: https://www.sciencedirect.com/science/article/pii/S1084804515002088
[67] V. Schmidt, K. Goyal, A. Joshi, B. Feld, L. Conell, N. Laskaris, D. Blank, J. Wilson, S. Friedler, and S. Luccioni, "CodeCarbon: Estimate and Track Carbon Emissions from Machine Learning Computing," 2021.
[68] M. Kumar, X. Zhang, L. Liu, Y. Wang, and W. Shi, "Energy-Efficient Machine Learning on the Edges," in *2020 IEEE International Parallel and Distributed Processing Symposium Workshops (IPDPSW)*, 2020, pp. 912–921.
[69] X. Dai, P. Zhang, B. Wu, H. Yin, F. Sun, Y. Wang, M. Dukhan, Y. Hu, Y. Wu, Y. Jia, P. Vajda, M. Uyttendaele, and N. K. Jha, "ChamNet: Towards Efficient Network Design through Platform-Aware Model Adaptation," *arXiv:1812.08934 [cs]*, Dec. 2018, arXiv: 1812.08934. [Online]. Available: http://arxiv.org/abs/1812.08934

[70] B. D. Rouhani, A. Mirhoseini, and F. Koushanfar, "DeLight: Adding Energy Dimension to Deep Neural Networks," in *Proceedings of the 2016 International Symposium on Low Power Electronics and Design, ser. ISLPED '16*. New York, NY, USA: Association for Computing Machinery, Aug. 2016, pp. 112–117. [Online]. Available: https://doi.org/10.1145/2934583.2934599

[71] X. Zhang, X. Zhou, M. Lin, and J. Sun, "ShuffleNet: An Extremely Efficient Convolutional Neural Network for Mobile Devices," in *Conference on Computer Vision and Pattern Recognition (CVPR)*, 2018, pp. 6848–6856.

[72] F. N. Iandola, S. Han, M. W. Moskewicz, K. Ashraf, W. J. Dally, and K. Keutzer, "Squeezenet: Alexnet-level accuracy with 50x fewer parameters and ¡0.5mb model size," 2016.

[73] J. Redmon and A. Farhadi, "Yolo9000: Better, faster, stronger," in *Proceedings of the IEEE Conference on Computer Vision and Pattern Recognition (CVPR)*, July 2017.

[74] W. Liu, D. Anguelov, D. Erhan, C. Szegedy, S. Reed, C.-Y. Fu, and A. C. Berg, "SSD: Single Shot MultiBox Detector," in *Computer Vision – ECCV 2016*, B. Leibe, J. Matas, N. Sebe, and M. Welling, Eds. Cham: Springer International Publishing, 2016, pp. 21–37.

[75] K. Greff, R. K. Srivastava, J. Koutník, B. R. Steunebrink, and J. Schmidhuber, "Lstm: A Search Space Odyssey," *IEEE Transactions on Neural Networks and Learning Systems*, vol. 28, no. 10, pp. 2222–2232, 2017.

[76] A. Kusupati, M. Singh, K. Bhatia, A. Kumar, P. Jain, and M. Varma, "FastGRNN: A Fast, Accurate, Stable and Tiny Kilobyte Sized Gated Recurrent Neural Network," *Advances in Neural Information Processing Systems*, vol. 31. Curran Associates, Inc., 2018. [Online]. Available: https://proceedings.neurips.cc/paper/2018/hash/ab013ca67cf2d50796b0c11d1b8bc95d-Abstract.html

[77] A. Tavanaei, M. Ghodrati, S. R. Kheradpisheh, T. Masquelier, and A. Maida, "Deep Learning in Spiking Neural Networks," *Neural Networks*, vol. 111, pp. 47–63, Mar. 2019. [Online]. Available: https://www.sciencedirect.com/science/article/pii/S0893608018303332

[78] Y. Venkatesha, Y. Kim, L. Tassiulas, and P. Panda, "Federated Learning with Spiking Neural Networks," *IEEE Transactions on Signal Processing*, vol. 69, pp. 6183–6194, 2021.

[79] B. Sudharsan, J. G. Breslin, and M. I. Ali, "Edge2Train: A Framework to Train Machine Learning Models (SVMs) On Resource-Constrained IoT Edge Devices," in *Proceedings of the 10th International Conference on the Internet of Things*. Malmo Sweden: ACM, Oct. 2020, pp. 1–8. [Online]. Available: https://dl.acm.org/doi/10.1145/3410992.3411014

[80] A. Kumar, S. Goyal, and M. Varma, "Resource-efficient Machine Learning in 2 KB RAM for the Internet of Things," p. 10.

[81] L. Torrey and J. Shavlik, "Transfer Learning," in *Handbook of research on machine learning applications and trends: algorithms, methods, and techniques*. IGI global, 2010, pp. 242–264.

[82] M. G. Arivazhagan, V. Aggarwal, A. K. Singh, and S. Choudhary, "Federated Learning with Personalization Layers," *CoRR*, vol. abs/1912.00818, 2019. [Online]. Available: http://arxiv.org/abs/1912.00818

[83] D. Li and J. Wang, "FedMD: Heterogenous Federated Learning via Model Distillation," *arXiv:1910.03581 [cs, stat]*, Oct. 2019, arXiv: 1910.03581. [Online]. Available: http://arxiv.org/abs/1910.03581

[84] H. Zhu, H. Zhang, and Y. Jin, "From Federated Learning to Federated Neural Architecture Search: A Survey," *Complex & Intelligent Systems*, vol. 7, no. 2, pp. 639–657, Apr. 2021. [Online]. Available: http://link.springer.com/10.1007/s40747-020-00247-z
[85] M. Xu, Y. Zhao, K. Bian, G. Huang, Q. Mei, and X. Liu, "Federated Neural Architecture Search," *arXiv:2002.06352 [cs]*, Jun. 2020, arXiv: 2002.06352. [Online]. Available: http://arxiv.org/abs/2002.06352
[86] H. Zhu and Y. Jin, "Real-time Federated Evolutionary Neural Architecture Search," *arXiv:2003.02793 [cs, stat]*, Mar. 2020, arXiv: 2003.02793. [Online]. Available: http://arxiv.org/abs/2003.02793
[87] C. Zhang, X. Yuan, Q. Zhang, G. Zhu, L. Cheng, and N. Zhang, "Towards Tailored Models on Private AIoT Devices: Federated Direct Neural Architecture Search," *IEEE Internet of Things Journal*, pp. 1–1, 2022, conference Name: IEEE Internet of Things Journal.
[88] C. He, M. Annavaram, and S. Avestimehr, "Towards Non-I.I.D. and Invisible Data with FedNAS: Federated Deep Learning via Neural Architecture Search," *arXiv:2004.08546 [cs, stat]*, Jan. 2021, arXiv: 2004.08546. [Online]. Available: http://arxiv.org/abs/2004.08546
[89] H. Liu, K. Simonyan, and Y. Yang, "DARTS: Differentiable Architecture Search," *arXiv:1806.09055 [cs, stat]*, Apr. 2019, arXiv: 1806.09055. [Online]. Available: http://arxiv.org/abs/1806.09055
[90] J. Xu and H. Wang, "Client Selection and Bandwidth Allocation in Wireless Federated Learning Networks: A Long-Term Perspective," *IEEE Transactions on Wireless Communications*, vol. 20, no. 2, pp. 1188–1200, 2021.
[91] K. Yang, T. Jiang, Y. Shi, and Z. Ding, "Federated Learning Via over-the-air Computation," *IEEE Transactions on Wireless Communications*, vol. 19, no. 3, pp. 2022–2035, 2020.
[92] T. Nishio and R. Yonetani, "Client Selection for Federated Learning with Heterogeneous Resources in Mobile Edge," *ICC 2019*, pp. 1–7, May 2019. [Online]. Available: http://arxiv.org/abs/1804.08333
[93] X. Liu, X. Qin, H. Chen, Y. Liu, B. Liu, and P. Zhang, "Age-Aware Communication Strategy in Federated Learning with Energy Harvesting Devices," in *2021 IEEE/CIC International Conference on Communications in China (ICCC)*, 2021, pp. 358–363.
[94] F. Lai, X. Zhu, H. V. Madhyastha, and M. Chowdhury, "Oort: Efficient Federated Learning via Guided Participant Selection," p. 18.
[95] C. Shen, J. Yang, and J. Xu, "On federated Learning with Energy Harvesting Clients," *arXiv preprint arXiv:2202.06105*, 2022.
[96] Y. Cui, K. Cao, G. Cao, M. Qiu, and T. Wei, "Client Scheduling and Resource Management for Efficient Training in Heterogeneous IOT-Edge Federated Learning," *IEEE Transactions on Computer-Aided Design of Integrated Circuits and Systems*, pp. 1–1, 2021.
[97] Z. Yang et al., "Energy Efficient Federated Learning over Wireless Communication Networks," *IEEE Transactions on Wireless Communications*, vol. 20, no. 3, pp. 1935–1949, 2021.
[98] Y. Cui, K. Cao, G. Cao, M. Qiu, and T. Wei, "Client Scheduling and Resource Management for Efficient Training in Heterogeneous IoT-Edge Federated Learning," *IEEE Transactions on Computer-Aided Design of Integrated Circuits and Systems*, pp. 1–1, 2021.

[99] V.-D. Nguyen, S. K. Sharma, T. X. Vu, S. Chatzinotas, and B. Ottersten, "Efficient Federated Learning Algorithm for Resource Allocation in Wireless IoT Networks," *IEEE Internet of Things Journal*, vol. 8, no. 5, pp. 3394–3409, Mar. 2021, conference Name: IEEE Internet of Things Journal.
[100] C. W. Zaw, S. R. Pandey, K. Kim, and C. S. Hong, "Energy-Aware Resource Management for Federated Learning in Multi-Access Edge Computing Systems," *IEEE Access*, vol. 9, pp. 34938–34950, 2021.
[101] X. Mo and J. Xu, "Energy-Efficient Federated Edge Learning with Joint Communication and Computation Design," *Journal of Communications and Information Networks*, vol. 6, no. 2, pp. 110–124, Jun. 2021, conference Name: Journal of Communications and Information Networks.
[102] X. Liu, X. Qin, H. Chen, Y. Liu, B. Liu, and P. Zhang, "Age-Aware Communication Strategy in Federated Learning with Energy Harvesting Devices," in *2021 IEEE/CIC International Conference on Communications in China (ICCC)*, Jul. 2021, pp. 358–363, ISSN: 2377-8644.
[103] R. Hamdi, M. Chen, A. B. Said, M. Qaraqe, and H. V. Poor, "Federated Learning over Energy Harvesting Wireless Networks," *IEEE Internet of Things Journal*, vol. 9, no. 1, pp. 92–103, Jan. 2022, conference Name: IEEE Internet of Things Journal.
[104] B. Guler and A. Yener, "Sustainable Federated Learning," *arXiv preprint arXiv:2102.11274*, 2021.
[105] J. Yang, X. Wu, and J. Wu, "Optimal Scheduling of Collaborative Sensing in Energy Harvesting Sensor Networks," *IEEE Journal on Selected Areas in Communications*, vol. 33, no. 3, pp. 512–523, Mar. 2015, conference Name: IEEE Journal on Selected Areas in Communications.
[106] A. Tak and S. Cherkaoui, "Federated Edge Learning: Design Issues and Challenges," *IEEE Network*, vol. 35, no. 2, pp. 252–258, 2021.
[107] A. Albaseer, M. Abdallah, A. Al-Fuqaha, and A. Erbad, "Threshold-Based Data Exclusion Approach for Energy-Efficient Federated Edge Learning," in *2021 IEEE International Conference on Communications Workshops (ICC Workshops)*, Jun. 2021, pp. 1–6, ISSN: 2694-2941.
[108] G. Drainakis, P. Pantazopoulos, K. V. Katsaros, V. Sourlas, A. Amditis, and D. I. Kaklamani, *From Centralized to Federated Learning: Exploring Performance and End-to-End Resource Consumption*, Social Science Research Network, Rochester, NY, SSRN Scholarly Paper ID 4026072, Feb. 2022. [Online]. Available: https://papers.ssrn.com/abstract=4026072
[109] Z. Ma, Y. Xu, H. Xu, Z. Meng, L. Huang, and Y. Xue, "Adaptive Batch Size for Federated Learning in Resource-constrained Edge Computing," *IEEE Transactions on Mobile Computing*, pp. 1–1, 2021.
[110] L. Li, D. Shi, R. Hou, H. Li, M. Pan, and Z. Han, "To Talk or to Work: Flexible Communication Compression for Energy Efficient Federated Learning over Heterogeneous Mobile Edge Devices," in *IEEE INFOCOM 2021 - IEEE Conference on Computer Communications*, May 2021, pp. 1–10, iSSN: 2641-9874.
[111] S. Lee, B. Islam, Y. Luo, and S. Nirjon, "Intermittent Learning: On-Device Machine Learning on Intermittently Powered System," *Proceedings of the ACM on Interactive, Mobile, Wearable and Ubiquitous Technologies*, vol. 3, no. 4, pp. 141:1–141:30, Dec. 2019. [Online]. https://doi.org/10.1145/3369837

[112] Z. Xu, Z. Yang, J. Xiong, J. Yang, and X. Chen, "ELFISH: Resource-Aware Federated Learning on Heterogeneous Edge Devices," *Proceedings of Machine Learning Research*, vol. 97, p. 634–643, 2019.
[113] S. Chen, C. Shen, L. Zhang, and Y. Tang, "Dynamic Aggregation for Heterogeneous Quantization in Federated Learning," *IEEE Transactions on Wireless Communications*, vol. 20, no. 10, pp. 6804–6819, Oct. 2021, conference Name: IEEE Transactions on Wireless Communications.
[114] E. Jeong, S. Oh, H. Kim, J. Park, M. Bennis, and S.-L. Kim, "Communication Efficient On-Device Machine Learning: Federated Distillation and Augmentation under Non-IID Private Data," *arXiv:1811.11479 [cs, stat]*, Nov. 2018, arXiv: 1811.11479. [Online]. Available: http://arxiv.org/abs/1811.11479
[115] D. Jhunjhunwala, A. Gadhikar, G. Joshi, and Y. C. Eldar, "Adaptive Quantization of Model Updates for Communication-Efficient Federated Learning," in *ICASSP 2021-2021 IEEE International Conference on Acoustics, Speech and Signal Processing (ICASSP)*, Jun. 2021, pp. 3110–3114, iSSN: 2379-190X.
[116] Y. Jiang, S. Wang, V. Valls, B. J. Ko, W.-H. Lee, K. K. Leung, and L. Tassiulas, "Model Pruning Enables Efficient Federated Learning on Edge Devices," *arXiv:1909.12326 [cs, stat]*, Oct. 2020, arXiv: 1909.12326. [Online]. Available: http://arxiv.org/abs/1909.12326
[117] S. Yu, P. Nguyen, A. Anwar, and A. Jannesari, "Adaptive Dynamic Pruning for Non-IID Federated Learning," *arXiv:2106.06921 [cs]*, Jun. 2021, arXiv: 2106.06921. [Online]. Available: http://arxiv.org/abs/2106.06921
[118] Y. Chen, X. Sun, and Y. Jin, "Communication-efficient Federated Deep Learning with Layerwise Asynchronous Model Update and Temporally Weighted Aggregation," *IEEE Transactions on Neural Networks and Learning Systems*, vol. 31, no. 10, pp. 4229–4238, 2020.
[119] P. Li, X. Huang, M. Pan, and R. Yu, "FedGreen: Federated Learning with Fine-Grained Gradient Compression for Green Mobile Edge Computing," *arXiv:2111.06146 [cs]*, Nov. 2021, arXiv: 2111.06146. [Online]. Available: http://arxiv.org/abs/2111.06146
[120] M. S. H. Abad, E. Ozfatura, D. Gunduz, and O. Ercetin, "Hierarchical Federated Learning ACROSS Heterogeneous Cellular Networks," in *ICASSP 2020 2020 IEEE International Conference on Acoustics, Speech and Signal Processing (ICASSP)*, May 2020, pp. 8866–8870, iSSN: 2379-190X.
[121] L. Liu, J. Zhang, S. Song, and K. B. Letaief, "Client-Edge-Cloud Hierarchical Federated Learning," in *ICC 2020 - 2020 IEEE International Conference on Communications (ICC)*, Jun. 2020, pp. 1–6, iSSN: 1938-1883.
[122] R. Ruby, H. Yang, F. A. P. de Figueiredo, T. Huynh-The, and K. Wu, "Energy Efficient Multi-Processor-Based Computation and Communication Resource Allocation in Two-Tier Federated Learning Networks," *IEEE Internet of Things Journal*, pp. 1–1, 2022, conference Name: IEEE Internet of Things Journal.
[123] C. Xie, S. Koyejo, and I. Gupta, "Asynchronous Federated Optimization," *arXiv:1903.03934 [cs]*, Sep. 2019. [Online]. Available: http://arxiv.org/abs/1903.03934
[124] G. Shi, L. Li, J. Wang, W. Chen, K. Ye, and C. Xu, "Hysync: Hybrid Federated Learning with Effective Synchronization," in *2020 IEEE 22nd International Conference on High Performance Computing and Communications; IEEE 18th*

International Conference on Smart City; IEEE 6th International Conference on Data Science and Systems (HPCC/SmartCity/DSS), 2020, pp. 628–633.

[125] J. Hao, Y. Zhao, and J. Zhang, "Time Efficient Federated Learning with Semiasynchronous Communication," in *2020 IEEE 26th International Conference on Parallel and Distributed Systems (ICPADS)*, 2020, pp. 156–163.

[126] K. Hsieh, A. Harlap, N. Vijaykumar, D. Konomis, G. R. Ganger, P. B. Gibbons, and O. Mutlu, "Gaia: {Geo-Distributed} Machine Learning Approaching {LAN} Speeds," 2017, pp. 629–647. [Online]. Available: https://www.usenix.org/conference/nsdi17/technical-sessions/presentation/hsieh

[127] K. Wang, R. Mathews, C. Kiddon, H. Eichner, F. Beaufays, and D. Ramage, "Federated Evaluation of On-device Personalization," *arXiv:1910.10252 [cs, stat]*, Oct. 2019, arXiv: 1910.10252. [Online]. Available: http://arxiv.org/abs/1910.10252

[128] V. Smith, C.-K. Chiang, M. Sanjabi, and A. Talwalkar, "Federated Multi-Task Learning," *arXiv:1705.10467 [cs, stat]*, May 2017, arXiv: 1705.10467 version: 1. [Online]. Available: http://arxiv.org/abs/1705.10467

[129] Y. Jiang, J. Konecnˇ Y. K. Rush, and S. Kannan, "Improving Federated Learning Personalization via Model Agnostic Meta Learning," *arXiv:1909.12488 [cs, stat]*, Sep. 2019, arXiv: 1909.12488. [Online]. Available: http://arxiv.org/abs/1909.12488

[130] H. Kim, J. Park, M. Bennis, and S.-L. Kim, "Blockchained On-Device Federated Learning," *IEEE Communications Letters*, vol. 24, no. 6, pp. 1279–1283, Jun. 2020, conference Name: IEEE Communications Letters.

[131] D. C. Nguyen, M. Ding, Q.-V. Pham, P. N. Pathirana, L. B. Le, A. Seneviratne, J. Li, D. Niyato, and H. V. Poor, "Federated Learning Meets Blockchain in Edge Computing: Opportunities and Challenges," *IEEE Internet of Things Journal*, vol. 8, no. 16, pp. 12806–12825, Aug. 2021, conference Name: IEEE Internet of Things Journal.

[132] S. R. Pokhrel and J. Choi, "Federated Learning with Blockchain for Autonomous Vehicles: Analysis and Design Challenges," *IEEE Transactions on Communications*, vol. 68, no. 8, pp. 4734–4746, Aug. 2020, conference Name: IEEE Transactions on Communications.

[133] X. Bao, C. Su, Y. Xiong, W. Huang, and Y. Hu, "FLChain: A Blockchain for Auditable Federated Learning with Trust and Incentive," Aug. 2019, pp. 151–159.

[134] K. Bonawitz, V. Ivanov, B. Kreuter, A. Marcedone, H. B. McMahan, S. Patel, D. Ramage, A. Segal, and K. Seth, "Practical secure aggregation for privacy-preserving machine learning," in *Proceedings of the 2017 ACM SIGSAC Conference on Computer and Communications Security, ser. CCS '17*. New York, NY, USA: Association for Computing Machinery, 2017, p. 1175–1191. [Online]. Available: https://doi.org/10.1145/3133956.3133982

Chapter 4

GREEN6G

Chameleon federated learning for energy-efficient network slicing in beyond 5G systems

Anurag Thantharate
University of Missouri Kansas City, Kansas City MI, USA

4.1 INTRODUCTION

The fifth-generation new radio (5G NR) is a wireless communication standard that supports various services with different requirements using logically isolated networks called 'slices.' These slices are designed to serve different users and services with unique characteristics, such as ultra-reliable low-latency communication, machine-type communication, and enhanced mobile broadband. Network slicing allows mobile operators to create multiple virtual networks within a single physical network, enabling them to offer customized solutions and innovative services to their clients using cloud and virtualized radio access networks. As the demand for emerging applications like autonomous driving and augmented reality grows, future 6G networks will need to be even faster, more reliable, and more adaptable, with improved machine learning algorithms considering diversity and privacy. These networks must also be scalable, reliable, low-latency, and energy-efficient, with efficient spectrum utilization.

Wireless network optimization using data collected from end users and core network elements is becoming increasingly important as the volume of data generated by end users and network elements grows. However, there are concerns about data privacy, as this data often contains confidential information. Distributed federated learning (FL) can address these concerns by allowing data to be trained locally on end-user devices rather than transmitted to a central server for network optimization modeling. This decentralized approach can help preserve user data and experience while allowing network optimization. The use of collaborative machine learning beyond 5G (B5G) systems, including network slicing systems, for heterogeneous device types has yet to be thoroughly investigated. One proposed solution is GREEN6G, a distributed learning framework for optimizing network slicing architecture. This framework seeks to carefully study the use of collaborative machine learning in B5G systems while preserving user data and experience. The current state of traditional Machine Learning frameworks is depicted in Figure 4.1, outlining how and where models are trained. Both

DOI: 10.1201/9781003230427-4

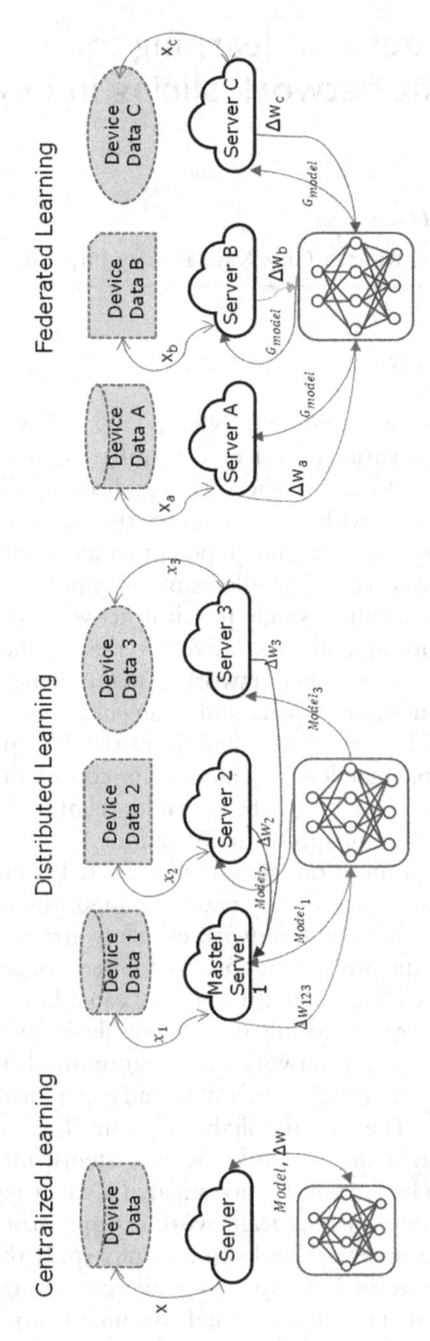

Figure 4.1 Different classes of machine learning.

centralized and distributed learning relies on a central server for data training and model creation, with end devices communicating with the server whenever they need to make a prediction. FL enables parallel learning across servers by allowing end devices to independently train a model using their local data on-device without sending sample data to the server for learning; in essence, FL brings the model to the data rather than sending it to the central server.

The term 'Chameleon' refers to the ability of a single machine learning model to adapt to a diverse set of end-user devices. In the context of network slicing, there may be different slices serving different types of devices and applications, such as smartphones and mobile broadband applications (Slice A), Internet-of-Things devices (Slice B), and latency-sensitive services like augmented and virtual reality (Slice C). Federated learning is training a shared model across devices by allowing them to train a model locally on their data and send the resulting model updates (weights, biases, and gradients) to a central server rather than raw data. Each slice receives the global model and modifies it through local training, and the updates are averaged to produce a single integrated model. This process is repeated multiple times to improve the accuracy of the model. This research aims to identify ways to improve machine learning algorithms' accuracy and computational efficiency in network slicing scenarios where data may be limited. The primary research direction for this chapter is how to implement FL in 5G network slicing architecture, as summarized below:

- when to upload the weights of which component (from which slices)
- how the cloud server organizes these weights to create a global model, and
- when to download/push the global model to end-devices

The rest of this chapter is structured as follows. Section 4.2 discusses the current state and challenges of traditional machine learning and our approach to FL in the B5G ecosystem and provides a background of the need for Green communication and how ML can help achieve a sustainable ecosystem. Section 4.3 discussed network slicing research and related work from other researchers. Section 4.4 explains the proposed algorithm and working methodology for GREEN6G, and the evaluation and plausible use cases are summarized in Section 4.5. A future course of GREEN6G is discussed in Section 4.6 as a conclusion.

4.2 NEED FOR GREEN COMMUNICATION AND ROLE OF MACHINE LEARNING IN B5G

Artificial Intelligence (AI) in B5G applies to many fields. It captures a variety of complex nonlinear or highly dynamic relationships between inputs and outputs that vary over time. Firstly, using AI in wireless communication and networking for content prediction, pattern recognition, regression, clustering, and failure detection attracts significant interest. In particular, computing and caching technology developments allow base stations to store and interpret human behavior for wireless communication. The two most important AI applications in wireless communication are: the first is the capacity of wireless communication systems to learn from data sets generated by users, environments, and network devices for prediction, inference, and interpretation of big data. Second, AI is implanted at the network's edge, the base station, and the device to enable self-organizing network operation. Self-organizing solutions for resource management, user connection, and data offloading rely heavily on edge intelligence. Existing machine learning is distinct from edge machine learning, where training data is not distributed uniformly to edge devices (eNodeB, mobile device). It costs compute time and energy to send device-generated data to the cloud for prediction. On the other hand, cloud-based edge machine learning maintains privacy because training data is stored locally on all devices. Training with much user-generated data, including sensitive samples, improves prediction accuracy.

The 5G wireless industry uses data generated by production networks and end-users to solve optimization problems such as capacity forecasting, traffic estimation, resource scheduling, and planned maintenance. Traditional machine learning and deep learning techniques are used to analyze this data generated by billions of connected devices and network elements. The industry has adopted a centralized machine learning approach, where the data and training are done on a central server. However, there is increasing interest in distributed and decentralized machine learning approaches that allow end-user devices to store data locally and train models without transmitting raw data to a central server. This can improve personal data security and enable machines to make dynamic, local decisions based on a trained model, enhancing the user experience, low-latency communications, and response times and reducing energy and computational costs [1].

Centralized learning is currently the most common ML workflow, in which models are trained on large central servers that update model parameters using large datasets. After the model is trained, it is distributed to the edge device via the cloud to make inferences at the edge. The edge device collects data, sends it to the central server for predictions, and then receives the results from the server. These factors, including bandwidth and server availability/downtime, plan data limits, and battery life, make it challenging to conclude by pushing the model to its limits. Machine learning algorithms use an iterative process to continuously learn from sample data and improve over time. However, in the context of 5G network slicing, which supports various use cases, more than a single generic model may be required to determine the optimal model parameters for all types of user equipment (UEs). For example, smartphones tend to require more data and bandwidth than Internet-of-Things (IoT) devices like sensors, which are constrained by power drain and other resource limitations. Augmented and virtual reality (AR/VR) applications and mission-critical services are more latency-sensitive. Additionally, many UEs are only occasionally active, such as IoT devices that periodically check in with the server and do not require constant high data connectivity. These factors must be considered when designing machine learning algorithms for 5G network slicing.

Similarly, a connected automobile, Vehicle-to-everything (V2X) UE, may experience multiple handoffs during mobility. However, the system can avoid these handovers if it recognizes the vehicle's regular route and permits it to connect to sites with the fewest possible handovers and consistent, dependable links along its route. GREEN6G is a distinctive federated model; it creates a decentralized distributed learning model for these unique slices and UE types, generating significantly different KPIs. In contrast, when it comes to Distributed Deep Reinforcement Learning (DRL) transforms the user into an agent with their DRL policy. Each Agent's policy can be trained independently or by downloading the model after data collection and training in the cloud. For independent movement, the amount of data arriving at each UE is small, and it is difficult to reflect the entire network situation; consequently, training may decrease performance. Furthermore, DRL training is time-consuming. FL may be used to resolve this issue which training as much as the UE can train and uploading the weights to the cloud. Cloud then generates a Global Optimized Model by averaging the collected weights. Periodically, the UE downloads the model created in this manner and acts with it until the next download [2].

Below are some of the advantages in the context of 5G systems and network slices with Federated Learning:

1. *Privacy:* By keeping the data on the device, federated learning ensures that sensitive information is not shared with a central server or third party. This is especially important in scenarios where data privacy regulations are strict.
2. *Data efficiency:* In network slicing, different slices may serve different applications with different requirements. Federated learning allows each device to contribute to training a model specific to its slice, making the most of the data available on the device.
3. *Communication efficiency:* Training a model on the device itself reduces the amount of data that needs to be transmitted over the network, improving communication efficiency and reducing the burden on the network.
4. *Robustness:* Federated learning allows devices to continue training a model even when they are offline or have limited connectivity, improving the robustness and reliability of the system.
5. *Scalability:* Federated learning allows the model to be trained on many devices, making it easier to scale the system to a more significant number of users.

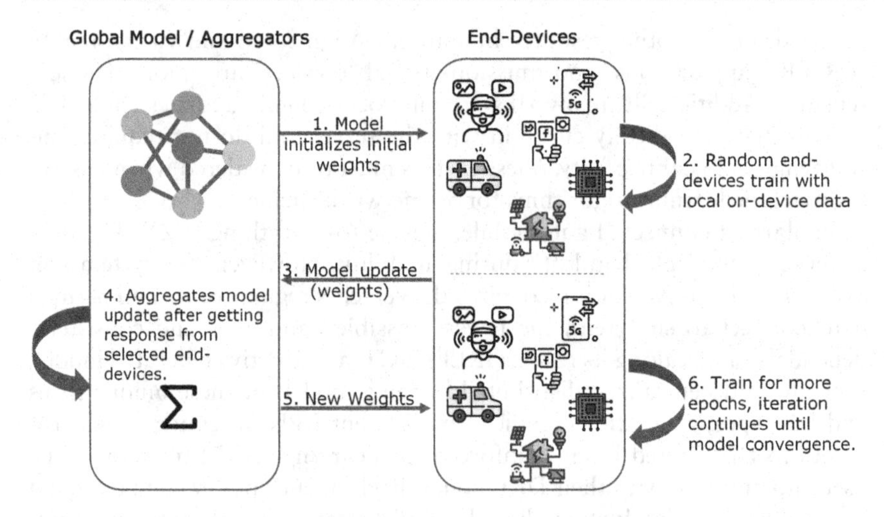

Figure 4.2 Chameleon federated framework for GREEN6G.

The purpose of FL is to utilize data from multiple locations. The general progression is as follows: Multiple devices in different locations (smartphones, IoT devices) train a local copy of a global ML model using local data. The updates trained on each device are then gathered on a global ML model on a central server, and a copy of the updated global model is sent back to each device to reflect all changes, as shown in Figure 4.2. Here is precisely how works:

1. Each device initially obtains a copy of the global ML model by either requesting the model from the server or sending the model to the device in advance. The model could be an early version with only random weights or a previously trained model.
2. The device collects local data to train a local copy of the model.
3. The device then transmits the delta to the cloud (such as weights and biases of the updated model). Currently, the delta represents the difference between the initial and trained models, so the training data are not transmitted externally.
4. Federated descent is achieved when a central server combines the deltas received from multiple devices or updates the global model using an algorithm like Federated Averaging. After incorporating the changes, the new, enhanced version of the global model is ready for inference. This model is then sent back to the edge device or utilized in the cloud for inference.

Due to environmental factors such as carbon emissions, there has been a rising concern about the amount of energy consumed by cellular networks. Moreover, the power consumption of cellular networks accounts for a substantial portion of operator Operating Expenses (OPEX) [3–5]. As a consequence, network energy efficiency becomes increasingly crucial. 3GPP has also started to drive efforts related to network energy conservation in TR 32.826, TR 36.887, TR 36.927, TR 37.816, and TR 21.866 specifications [6].

Despite FL capabilities, its high energy consumption prevents its use in 5G mobile devices with limited battery life. On-device processing and local model updates are energy-intensive and drain mobile device batteries. First, the system must increase energy efficiency to extend the life of mobile devices during training, where energy is dissipated by local computing and wireless communications. GREEN6G aims to enable effective and efficient local training on 5G mobile devices while minimizing communications and computing energy consumption. Several advanced techniques, including gradient descent, gradient quantization, weight quantization, model pruning, and dynamic batch sizing, are incorporated into this study to understand how to reduce the energy consumption of FL over 5G mobile devices. Our studies show that gradient descent, gradient quantization, and dynamic batch sizing conserve energy for communication, while weight quantization and model pruning reduce energy consumption for computing. These techniques should be combined to enable energy-efficient FL on 5G mobile devices.

As all 5G mobile devices train multiple times and transmit local model updates to the central aggregator over wireless transmissions, it consumes energy. The channel conditions may fluctuate, which can consume more power due to poor radio conditions, low-bandwidth issues, and retransmissions. The total energy consumed by each device during training for wireless transmissions is the product of the required number of global communication rounds multiplied by the energy consumption of a single round. Energy consumption per round depends on transmission power, rate, and model size. So, there are two ways to conserve energy during the communication portion of the training process: reducing the required rounds of global communications. After several local computing iterations, global synchronizations can be performed to minimize communication frequency (federated averaging). Second, by increasing the size of training batches, the system can reduce the number of rounds of global communication. Because FL convergence requires constant data computation, these two methods reduce the number of communication rounds by increasing the amount of data computation in each round. Model descent and quantization can decrease the local model size that must be transmitted, thereby conserving communication energy in each round.

4.3 RELATED WORK

FL is a new paradigm in wireless communication and is under-explored for a network function virtualization concept like network slicing. Industries are currently investigating ways to optimize Self-supervised Learning,

Graph Neural networks, and Efficient Network Architecture Design, focusing on network compression and lightweight model design. It is challenging to build a sizeable mobile model with limited computational resources. A light, powerful mobile model (on-device) is an exciting area of research. Other areas, such as Network Pruning, Model Quantization, Knowledge Distillation, Neural Network Architecture Search, and Dynamic Inference, are also under research studies.

The GREEN6G framework is the first contribution of distributed learning techniques that takes both UE and RAN (network) elements into account for the heterogeneous UE types and the diverse traffic for network slicing in 5G systems, considering compute energy, time, and cost. DeepSlice [7] and Secure5G [8] are approaches to studying network slices in 5G systems by applying machine and deep learning techniques. Authors have demonstrated network slice selection for all UE types, including unknown devices, load balancing techniques in case of slice failure, and security of these slices in case of a DDoS attack. The authors in [9, 10] presented the integration of federated and 6G learning and provided possible federated 6G learning applications. The authors have described the critical technical challenges, appropriate FL methods, and problems in the 6G communication context for future research into FL. An FL-based content caching algorithm, which considers user privacy while content caching on a real-world dataset is studied by Z. Yu et al. in [11]. The authors in [12] produced a research survey that discussed and summarized the current state and future directions of deep learning along with challenges. G. Wang et al. presented a hybrid network slice reconfiguration framework with a resource reservation mechanism to accommodate future traffic demand increases instead of reconfiguration [13]. The issues related to the wireless channel, like latency and energy consumption while training a machine learning model on end devices, are discussed in [14], keeping FL in context. Following the same problem statement, the authors in [15] have discussed joint device scheduling and resource allocation policy considering the training time budget for latency-constrained wireless networks. Implementing energy-intensive ML-based solutions in an extensive networking system is a significant challenge for future sustainability. However, federated learning's energy consumption has yet to be discovered.

The authors in [16] examine the subject of service provider revenue maximization in network slicing, intending to provide customized services for various user classes and diverse requirements. Precisely, the network is sliced based on the end user's perspective, which is examined at multiple levels. Consumers' demand forecasting for service classes is pursued by implementing a federated learning framework in which end users serve as customers. In [17], authors examined the end-to-end latency bound with given traffic demand and resources using stochastic network calculus (SNC). Then, we offer a method for determining the optimal allocation of resources, given a particular traffic distribution and end-to-end delay constraint. In addition, the authors study the range of traffic demands that a network slice can serve and design a learning-based dynamic network slice resizing method that can significantly lower the overall resizing cost while guaranteeing the quality of service. The authors in [18] proposed using a Federated Learning model, which entails storing raw data where it is created and sending only users' locally trained models to a centralized entity for aggregation for predicting the service-oriented KPIs of slices. The primary goal in [19] is to reduce the total offloading time and network resource consumption by optimizing compute offloading, resource allocation, and service caching placement. Authors used FL to train the 2Ts-DRL model in a distributed way to protect the data privacy of edge devices. Their Simulation findings confirm the efficacy of both 2Ts-DRL and FL in the I-UDEC framework and demonstrate that our suggested method can reduce task execution time by up to 31.87 percent. Deep reinforcement learning was used to optimize training time and energy consumption [20]. Yang et al. [21] proposed a low-complexity iterative algorithm to reduce energy consumption and latency. Dinh et al. [22] studied convergence and energy consumption. Sun et al. [23] proposed an online energy-aware dynamic edge server scheduling policy to maximize the average number of edge servers in a single iteration.

[24] This paper examines network level and radio access energy efficiency issues. BS (architecture, switching techniques, interference reduction, and deployment of small cells), network (resource sharing and allocation), SDN level (energy monitoring and management), and machine learning have been investigated concerning 5G energy conservation. The energy efficiency issues associated with the edge network portion are not addressed, nor are all machine learning technologies. Regarding resource allocation, network planning, energy harvesting, and hardware, the author discussed the energy efficiency aspect of 5G technology in [25]. The importance of network planning to achieve greater energy efficiency, offloading techniques, dense network deployment, and network benefit-cost ratio to achieve Quality of Service (QoS) has been emphasized. The research is effectively summed up while highlighting the network planning and hardware issue that impacts energy efficiency. However, the latest trend of cloud and virtualization, which has significant energy efficiency benefits, still needs to be addressed. In [26], the author discussed energy efficiency factors based on capacity expansion. Spectrum efficiency, spectrum reuse, and spectrum resources are the basis for this study. The author discussed green techniques in a spectrum context. Regardless of the significance of machine learning to energy efficiency, it effectively solves large-scale network problems.

Recent work [27] presents a Chameleon FL model, FED6G, for network slicing in 5G and beyond systems to tackle complex resource optimization problems without collecting sensitive end-device data. The evaluation results demonstrate a 39% improvement in Mean Squared Error (MSE), a 46% improvement in model accuracy, and a 23% reduction in energy cost while training the proposed FED6G neural network model vs. the conventional deep learning neural network model. The study in [28] presents Balanced5G, a data-driven traffic steering system that takes proactive actions during the HO stage to route traffic equally among various frequencies (low, mid, and high). Balanced5G guarantees that UEs (fixed or mobile) do not select a frequency purely based on the most robust signal strength but also consider other network essential parameters such as network load, active network slices, and the type of service for which the UE is demanding resources. Research simulation findings indicate that the suggested traffic steering strategy achieves forecasts that are 19.5% more accurate than conventional.

4.4 CHAMELEON FEDERATED LEARNING – GREEN6G FRAMEWORK

Federated learning (FL) encompasses two types. The first is cross-device learning, which involves training a model on multiple devices and collecting configurations from a wide range of devices, ranging from a few to thousands or even millions. This approach of cross-federated learning across numerous devices significantly improves the user experience. In this scenario, each device stores its historical data and performs local model training. The trained delta, representing the model updates, is then transmitted to a central server, where it is incorporated into the global model.

The second type is cross-silo federated learning. Comparable to the first type, but operating between organizations instead of personal smartphones or Internet-of-Things (IoT) devices, cross-silo federated learning is valuable for advancing machine learning (ML) models with large datasets. Let's consider an example where multiple organizations, such as hospitals, collaborate to enhance a machine learning model for diagnosing a rare disease. Due to the rarity of the disease and concerns related to data privacy, the availability of training data is scarce. However, cross-silo federated learning enables these organizations to leverage their respective data to create a model delta, benefiting all parties involved, including the hospitals. This approach allows them to collectively contribute their data while addressing data privacy concerns and overcoming the challenge of limited training data for rare disease. GREEN6G adapts the custom version of cross-device, and the model flow comprises four steps:

Step I: GREEN6G framework initializes the Global model (G_{CFL}) a random set of devices across slices A (eMBB), B (mIoT), and C (URLLC). Selected clients (x_{Ai}, x_{Bi}, x_{Ci}) receive the model and improve its parameter locally (on-device) using ground-truth local data.

Step II: Given the query from the Global model, each selected device computes a reply based on its local data, i.e., $w_{Ai} = (w_i{}^* x_{Ai})$, $w_{Bi} = (w_i{}^* x_{Bi})$, $w_{Ci} = (w_i{}^* x_{Ci})$. After computing, end devices independently forward the optimized model parameters updates (weights, gradients - w_{Ai}, w_{Bi}, w_{Ci}) to the slice server for aggregation.

Step III: The aggregation server averages the weight (Δw) from all end device types to improve the Global model and send the updated model back to devices again across slices.

Step IV: The iteration process continues until the model parameters converge as predetermined criteria.

In this section, the detailed step-by-step working of the proposed GREEN6G as shown in Figure 4.3:

a. GREEN6G framework initializes the global model (G_{CFL}) to random 'i' end-UEs. Network operators can deploy many numbers of slices. Still, this chapter is considering three standard slices, i.e., eMBB (Slice A), mIoT (Slice B), and URLLC (Slice C) for GREEN6G based on standardized 3GPP SST values, which further sends the aggregated updates for aggregation. End devices have access to local data, which will be used to train received models locally.
b. At the beginning of each communication round, a subset of random end-devices is chosen (can be operator defined). End devices receive the initial global model to update their local model (i.e., gradients,

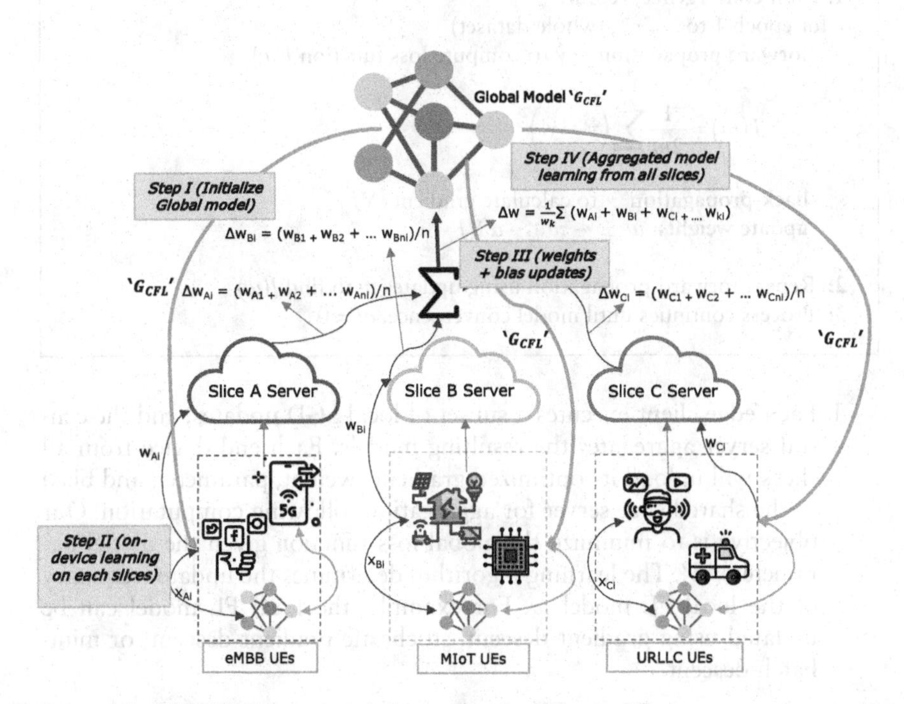

Figure 4.3 GREEN6G framework for network slicing in 5G.

weights) and start training locally using local data on-device. Sampled users from each slice can be written as:

$$\text{Cap}\, X_{Ai} = \left[x_{Ai1,\ldots,}\, x_{AiK_i}\right] \tag{4.1}$$

$$X_{Bi} = \left[x_{Bi1,\ldots,}\, x_{BiK_i}\right] \tag{4.2}$$

$$X_{Ci} = \left[x_{Ci1,\ldots,}\, x_{CiK_i}\right] \tag{4.3}$$

where K_i is the sample collected by users across Slice A, B, and C.

c. To develop an efficient model for training data, the system looks for a loss function that quantifies how far the learned model distribution deviates from the empirical distribution. The model executes one or more iterations (batches, epochs) of a learning algorithm employing Gradient Descent (GD), mini-batch GD, or Stochastic Gradient Descent (SGD) to obtain a model for the training data that minimizes a loss function.

Algorithm 1(a) Batch Gradient Descent

1: Until convergence, repeat:
for epoch 1 to .... n^{th}, (whole dataset)
forward propagation → $\hat{y}$ to compute loss function $J(w)$

$$J(w) = \frac{1}{2m}\sum_{i=1}^{m}\left(\hat{y}^i - y^i\right)^2$$

back-propagation → to calculate gradients ∇J
update weights; $w_{new} \leftarrow w_{old} - \alpha\, \nabla J$

2: Repeat forward propagation using updates w to find $J(w)$
3: Process continues until model convergence, $\partial J = 0$

I. Each edge client executes a subset of local SGD updates, and the central server aggregates the resulting models. Each end device from all slices will upload its optimized gradient (weight, parameter, and bias) to the shared slice server for aggregation following computation. Our objective is to minimize the global loss function given the model parameters 'w'. The learning algorithm determines the update frequency of the local FL model w. For example; the local FL model can be updated using gradient descent, stochastic gradient descent, or mini-batch descent.

Algorithm 1(b) Mini-Batch Stochatic Gradient Descent

1: Until convergence, repeat:
for epoch 1 to .... n^{th},
for $i = 1: m$ #random shuffle data in batch size
forward propagation → $\hat{y}$ to compute loss function $J(w)$

$$J(w) = \frac{1}{m}\sum_{i=1}^{m}(\hat{y}^i - y^i)$$

back-propagation → to calculate gradients $\nabla J = \frac{\delta \mathcal{L}}{\delta_w}$

update weights; $w_{new} \leftarrow w_{old} - \alpha \nabla J$

2: Repeat forward propagation using updates w to find $J(w)$
3: Process continues until model convergence, $\partial J = 0$

II. The global model is updated with a newly optimized parameter and sent back to another set of endpoints for additional optimization and iteration. It is important to note that, due to slicing and varying end-device participation, the end devices are likely to hold an unequal number of training samples massively distributed across slices. As a result, each local model in GREEN6G is weighted by the number of available training samples to aggregate the local models, as shown in the following expression:

$$\min \frac{1}{n}\sum_{i=1}^{K}\sum_{n=1}^{n_i} f\{(w_{Ai}, x_{Ai}, y_{Ai}), (w_{Bi}, x_{Bi}, y_{Bi}), (w_{Ci}, x_{Ci}, y_{Ci})\} \quad (4.4)$$

Where:

y_{Ai}, y_{Bi}, y_{Ci} is the output vector
w_{Ai}, w_{Bi}, w_{Ci} is obtained by training inputs $x_{Ai}, x_{Bi}, x_{Ci}, y_{Ai}, y_{Bi}, y_{Ci}$
k is the number of sampled participants for each round.
n_i is a training sample from all participants.
W_{t+1}^{k} is a local model parameter of end-device 'i'

III. The iteration process continues; the global server selects the aggregated optimized weights and updates its global model; the server then deploys the 'new updated global model' to the subset of devices in a non-IID fashion, and the process continues for predictions and further training until the model parameters converge according to a predetermined convergence criterion. The updated global model G_{CFL} after each iteration is as follows:

$$G_{CFL} = \frac{\sum_{i=1}^{K} n_i w_i}{n} \tag{4.5}$$

Machine learning algorithms often involve optimizing and minimizing a cost function by iteratively adjusting model parameters to fit a dataset. There are two main ways to train a model: incrementally and in bulk. In incremental training, the model is updated after processing each input, while in bulk training, the model is updated after processing all inputs in a batch. Gradient descent is a commonly used optimization algorithm that can be used in batch, stochastic, and minibatch training. It involves updating the model parameters based on the gradient of the cost function with respect to the parameters. However, it can be slow when applied to large or evolving datasets. Stochastic gradient descent (SGD) is a variant of gradient descent that involves updating the model parameters based on a single training sample at a time rather than the entire dataset. This can make SGD more efficient and faster than traditional gradient descent when applied to large datasets.

In addition, the path to minima is noisier (higher parameter update variance) due to randomness than GD. To make GREEN6G more productive, the model has implemented the Minibatch features. After each iteration, the training examples (samples from end devices) are randomly shuffled, and the cost gradient of selected data samples is applied. This reduces the cost of calculations and the variance and volatility of parameter updates. The proposed GREEN6G code and its pseudo code are as follows:

Algorithm 2(a): CFL Server Initialization

1: **procedure** GLOBALMODEL (G)
2: **iteration1:**
3: **initialize** G across Slice A, B and C;
4: **for all** (x_{Ai}), (x_{Bi}), (x_{Ci}) in the Slices A, B, and C **do**
5: $w_{t+1} \leftarrow$ ClientUpdate $(x_{Ai}, \Delta w_{Ai})$, $(x_{Bi}, \Delta w_{Bi})$, $(x_{Ci}, \Delta w_{Ci})$
6: Server computes new global model

$$w_{t+1} \leftarrow \sum_{i=1}^{k} \frac{n_i}{n} W_{t+1}^{k}$$

7: **end for**
8: **end procedure**

Algorithm 2(b): CFL Client Update

1: **procedure** ClientUpdate (x_{Ai}, Δw_{Ai}), (x_{Bi} Δw_{Bi}), (x_{Ci}, Δw_{Ci})
2: **for** each epoch (E) **do**
  for each mini-batch (B) **do**
    select subset of random client (x_{Ai}), (x_{Bi}), (x_{Ci}) ;
3: **initialize** G;
4: **for** all (x_{Ai}), (x_{Bi}), (x_{Ci}) in the Slices A, B, and C **do**
5: w computes average gradient ∇J with SGD
  update weights, $w_{t+1} \leftarrow w_t - \alpha \nabla J$;
7: **end for**
8: **return** (Δw_{Ai}), (Δw_{Bi}), (Δw_{Ci}) to server for aggregation $\rightarrow w$
9: **end procedure**

4.5 GREEN6G EVALUATION AND LOAD PREDICTION

The 5G wireless industry is using data generated by production networks and end-users to solve optimization problems such as capacity forecasting, traffic estimation, resource scheduling, and planned maintenance. These problems are tackled using traditional machine learning and deep learning techniques applied to data generated by billions of connected devices and network elements. Researchers are exploring distributed and decentralized machine learning approaches that allow end-user devices to store data locally and improve personal data security by eliminating raw data transmission to a central server. The GREEN6G framework is one proposed solution that addresses optimization problems in next-generation wireless networks, particularly in network slicing, multi-access edge computing, and cloud-based virtualized radio access network architecture. These environments require low latency, high bandwidth, and high quality of service and may involve highly heterogeneous and dense devices and data.

A cognitive network requires systematically modifying and configuring network resources based on traffic and bandwidth demands for varying geographic, temporal, and usage requirements. Adopting network slicing beyond 5G networks necessitates the execution of these projections at the Slice level. Predictions from the CFL model can aid in comprehending capacity and traffic demand, enabling network operators to configure slice automation based on anticipated demand, thereby preventing excessive overprovisioning or under-provisioning of resources. It is only sometimes feasible or practical to accurately compute the forecasted values using a small subset; therefore, the proposed model opted to 'estimate' rather than predict the network load and forecasted traffic with CFL.

This chapter evaluated GREEN6G by slicing networks with actual and artificial data by replicating eMBB, mIoT, and URLLC-like services on live 5G (Sub-6 and mmWave) networks using commercially available 5G devices to generate augmented data for slicing [7]. Modeled GREEN6G dataset [29] will help close the gap in machine learning modeling, particularly for Beyond 5G and network slicing systems-focused research communities. Using sequential modeling, the system splits the observed (actual) value dataset into train and test subsets to train our neural model. Mean Absolute Error (MAE), Mean Square Error (MSE), and Root Mean Square Error (RMSE) have been used to indicate the accuracy of CFL model prediction. MAE loss is advantageous in our data due to the fluctuating network load at various times. It is represented by the difference between the original and expected values derived from the absolute difference distributed across the data set. Our model's Mean Absolute Error and Mean Absolute Percentage Error (MAPE) can be expressed as:

$$MAE_A = \frac{1}{n}\sum_{i=1}^{n}\left(y_A - \hat{y}_A\right) \tag{4.6}$$

$$MAE_B = \frac{1}{n}\sum_{i=1}^{n}\left(y_B - \hat{y}_B\right) \tag{4.7}$$

$$MAE_C = \frac{1}{n}\sum_{i=1}^{n}\left(y_C - \hat{y}_C\right) \tag{4.8}$$

$$MSE_A = \frac{1}{n}\sum_{i=1}^{n}\left(y_A - \hat{y}_A\right)^2 \tag{4.9}$$

$$MSE_B = \frac{1}{n}\sum_{i=1}^{n}\left(y_B - \hat{y}_B\right)^2 \tag{4.10}$$

$$MSE_C = \frac{1}{n}\sum_{i=1}^{n}\left(y_C - \hat{y}_C\right)^2 \tag{4.11}$$

$$RMSE_A = \left[\frac{1}{n}\sum_{i=1}^{n}\left(y_A - \hat{y}_A\right)^2\right]^{\frac{1}{2}} \tag{4.12}$$

$$RMSE_B = \left[\frac{1}{n}\sum_{i=1}^{n}\left(y_B - \hat{y}_B\right)^2\right]^{\frac{1}{2}} \tag{4.13}$$

$$RMSE_C = \left[\frac{1}{n} \sum_{i=1}^{n} \left(y_C - \hat{y}_C \right)^2 \right]^{\frac{1}{2}} \tag{4.14}$$

R^2 (also known as the Coefficient of Determination) measures how close the estimated data are to the fitted regression line. The Coefficient of Determination tells us how many data points are utilized in the output of the regression line. R-squared can take any value between 0 and 1; the closer to 1, the more accurately it predicts the dependent variable. In initial model, the system obtained R^2 values of 0.972 for Slice A, 0.984 for Slice B, and 0.989 for Slice C, indicating that the fit explains approximately 97.2%, 98.4%, and 98.9% of the total variation in the data about the mean for Slice A, B, and C, respectively. The blue 'x' represents the predicted data in Figure 4.4 (a), (b), and (c), while the red line represents our model's prediction for each slice.

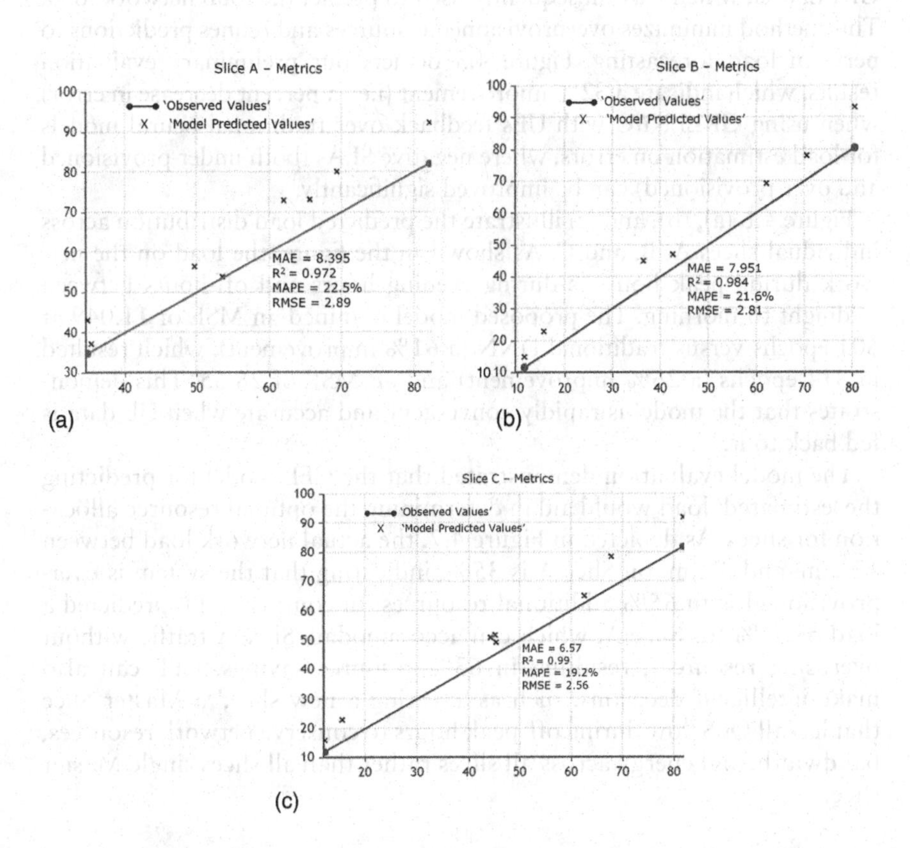

Figure 4.4 (a, b, and c) – model metrics and evaluation of CFL.

$$R_A^2 = 1 - \frac{\Sigma(y_A - \hat{y}_A)^2}{\Sigma(y_A - \bar{y}_A)^2} \tag{4.15}$$

$$R_B^2 = 1 - \frac{\Sigma(y_B - \hat{y}_B)^2}{\Sigma(y_B - \bar{y}_B)^2} \tag{4.16}$$

$$R_C^2 = 1 - \frac{\Sigma(y_C - \hat{y}_C)^2}{\Sigma(y_C - \bar{y}_C)^2} \tag{4.17}$$

Our model estimated the loss using regression's 'Mean Squared Error' loss function so that the model can adjust its weights after each iteration and optimize using the GD learning optimizer. Our evaluation method includes metrics such as the training error rate between traditional DNN and GREEN6G, which was subsequently used to predict the total network load. This method minimizes overprovisioned resources and refines predictions to perform load forecasting. Figure 4.5 depicts our preliminary evaluation results, which indicate a 32% improvement (i.e., a percent decrease in error) when using GREEN6G with UEs feedback over traditional neural models for load estimation on errors, where negative SLAs (both under-provisioned and over-provisioned) can be improved significantly.

Figure 4.6 (a), (b), and (c) illustrate the predicted load distribution across individual slices A, B, and C. As shown in the figure, the load on the network during peak hours is during evening hours and off-hours between midnight to morning. The proposed model obtained an MSE of 11.049 at 601 epochs versus traditional DNN (a 61% improvement), which resulted in 814 epochs (a 26% improvement) and an MSE of 28.65. This demonstrates that the model is rapidly convergent and accurate when UE data is fed back to it.

The model evaluation demonstrated that the CFL model for predicting the 'estimated' load would aid in determining the optimal resource allocation for slices. As depicted in Figure 4.7, the actual network load between 12 a.m. and 3 a.m. on Slice A is 35%, indicating that the system is over-provisioned with 65% additional resources. In contrast, CFL predicted a load of 37% for Slice A, which can accommodate Slice A traffic without overusing resources, resulting in 63% resource savings. CFL can also make intelligent decisions, such as 'creating a new slice,' a Master Slice that has all QoS flow during off-peak hours to conserve network resources, bandwidth, and energy across all slices rather than all slices single Master slice.

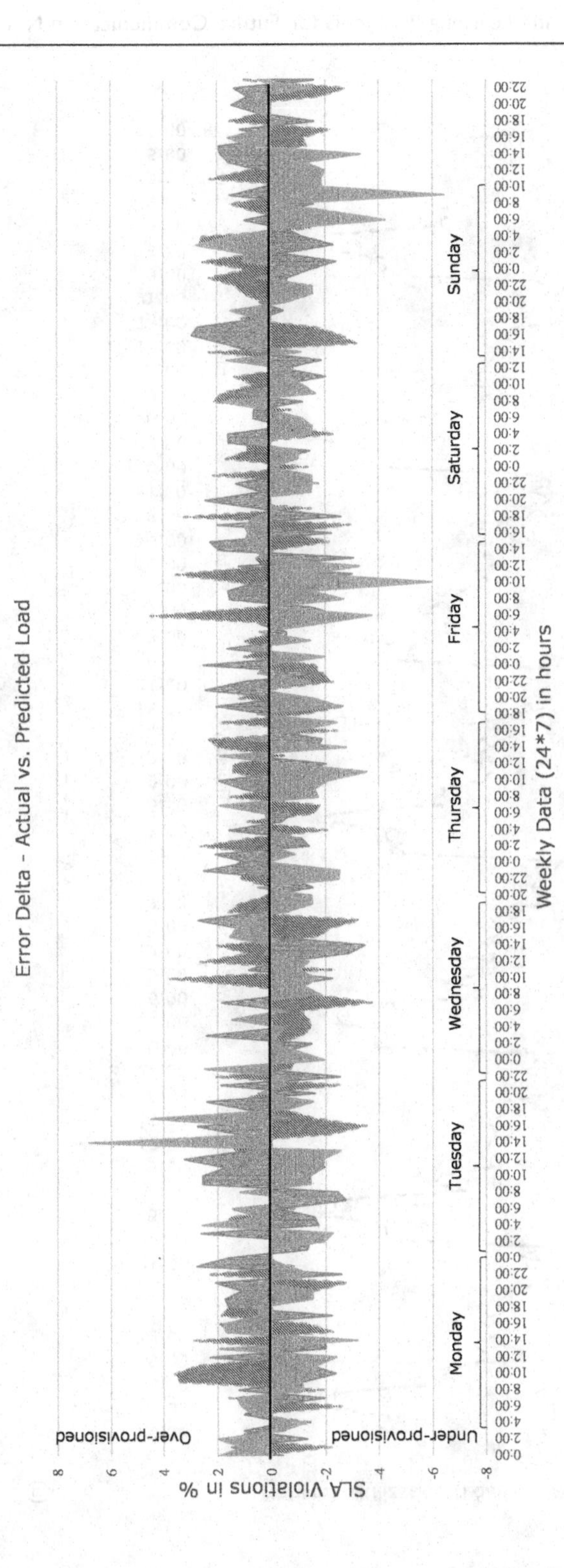

Figure 4.5 Training error between traditional DNN vs. GREEN6G.

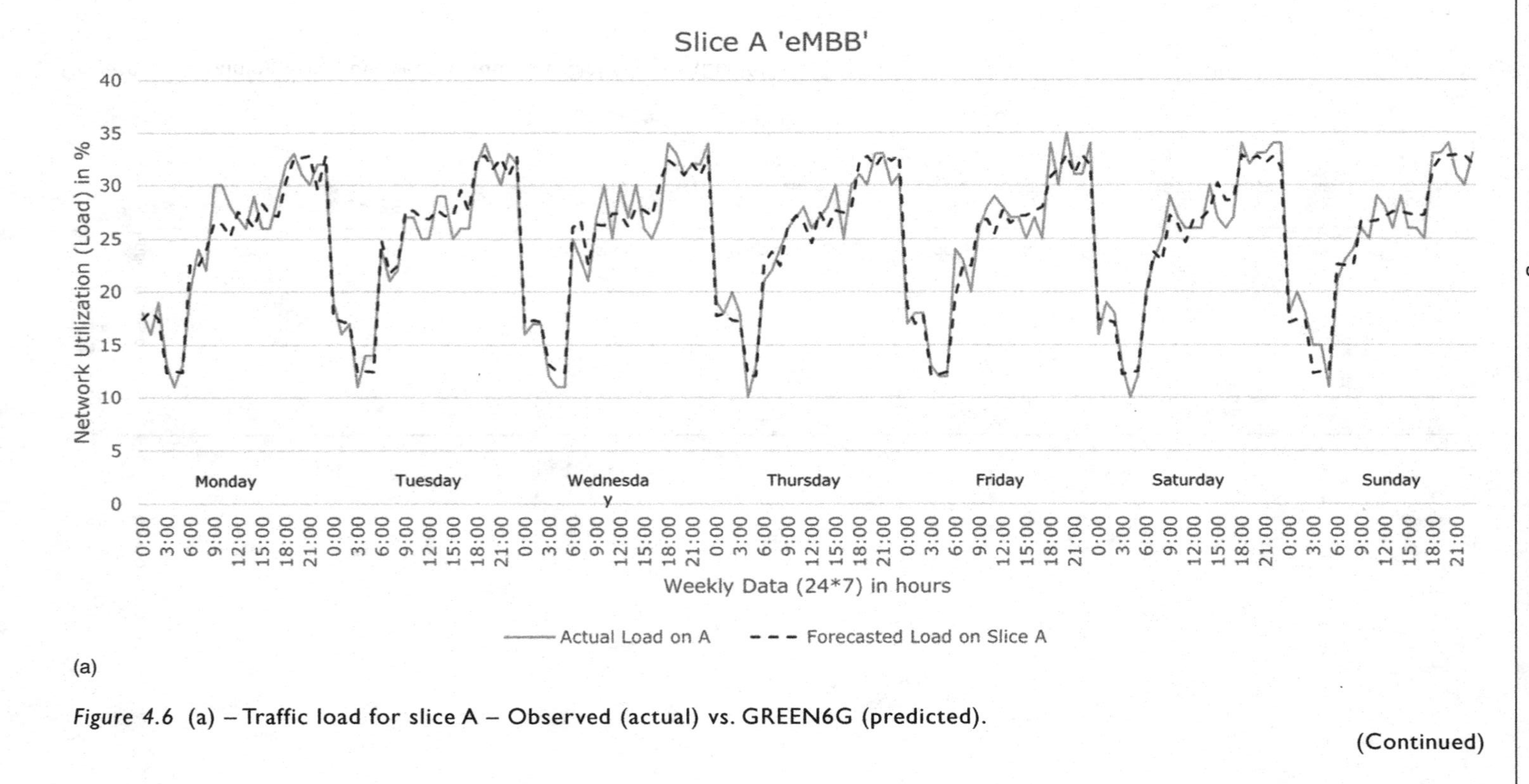

Figure 4.6 (a) – Traffic load for slice A – Observed (actual) vs. GREEN6G (predicted).

(Continued)

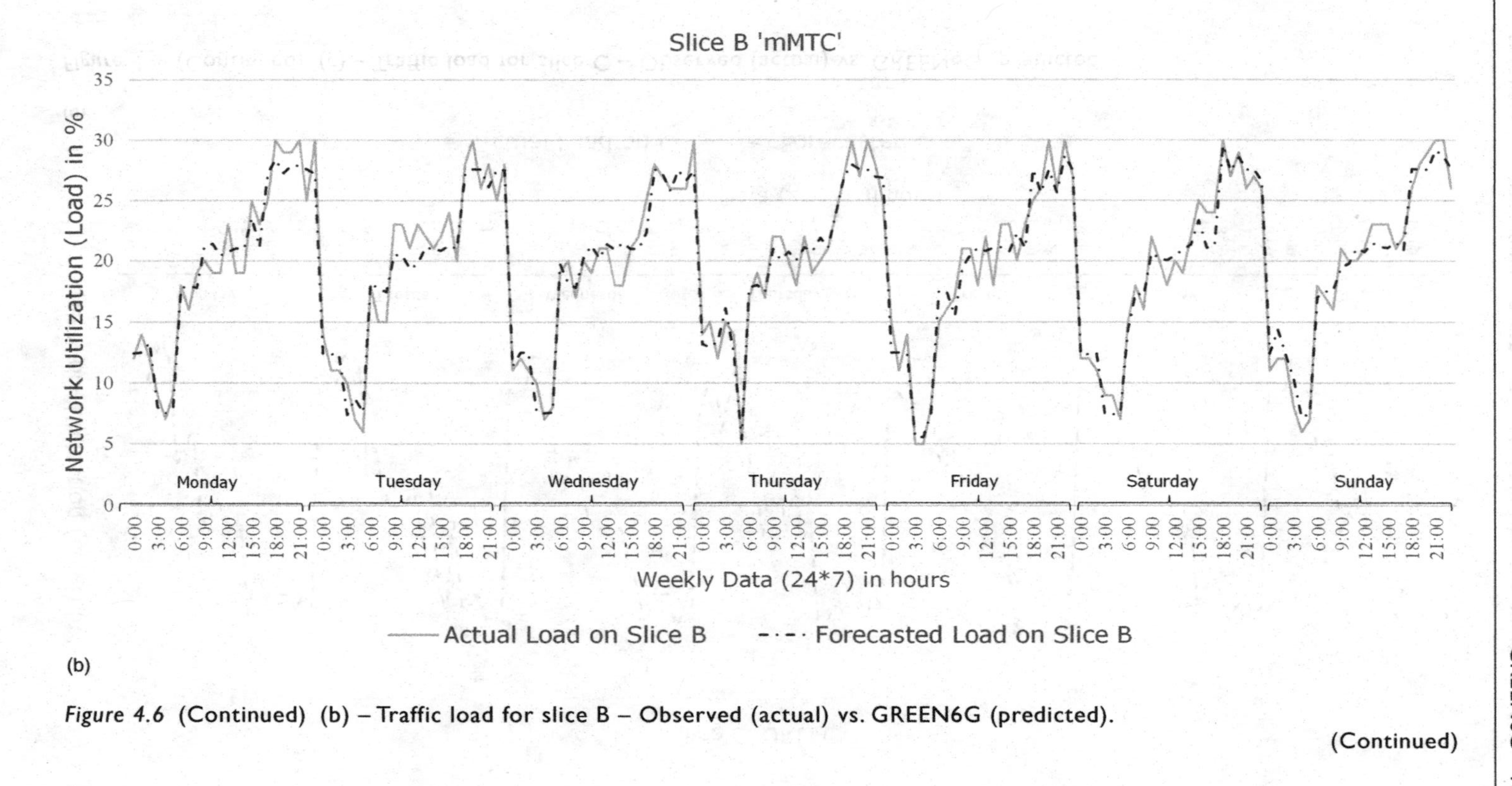

Figure 4.6 (Continued) (b) – Traffic load for slice B – Observed (actual) vs. GREEN6G (predicted).

(Continued)

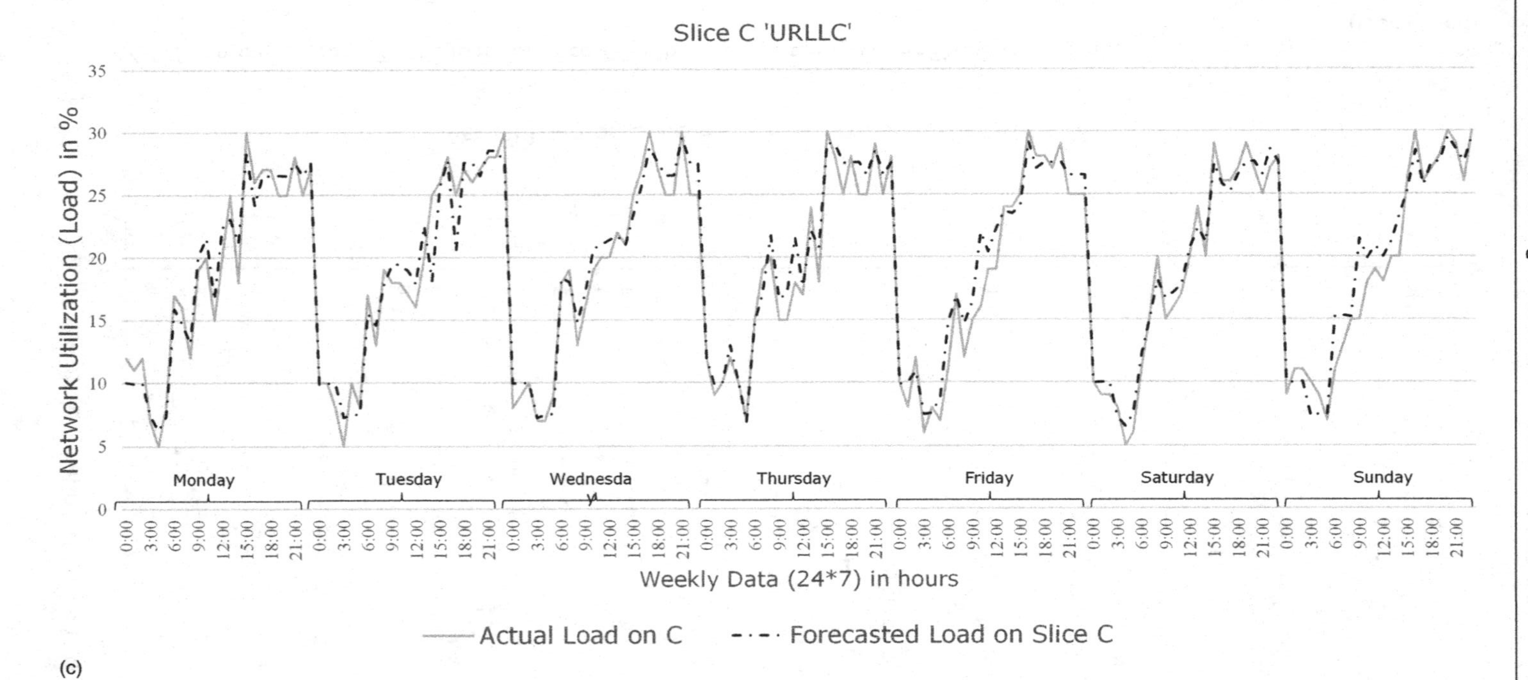

Figure 4.6 (Continued) (c) – Traffic load for slice C – Observed (actual) vs. GREEN6G (predicted).

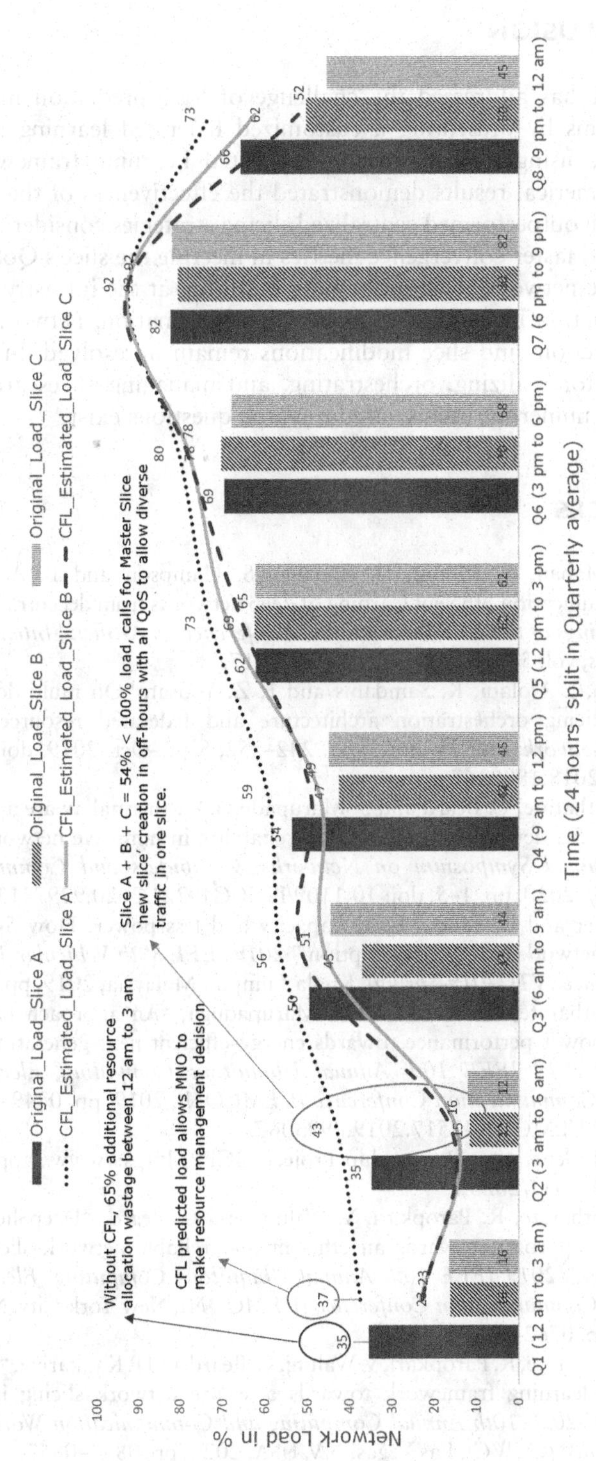

Figure 4.7 **Traffic load and resource utilization - observed (actual) vs. GREEN6G (predicted).**

4.6 CONCLUSION

This chapter has addressed the challenge of load prediction in network slicing systems by presenting a customized federated learning approach, 'GREEN6G,' using the Chameleon Federated Learning framework. The obtained numerical results demonstrated the effectiveness of the proposed model, which outperformed centralized slicing strategies considering energy, compute, and faster convergence metrics in meeting the slice's QoS requirements. While network slicing is rapidly maturing at the infrastructure and network function layers, issues such as load estimation, network reliability, slice selection, and slice modifications remain unresolved. In addition, approaches for realizing, orchestrating, and managing slices are in their infancy, and numerous unanswered research questions exist.

REFERENCES

[1] B. McMahan, E. Moore, D. Ramage, S. Hampson and B. A. Y. Arcas, "Communication-efficient learning of deep networks from decentralized data," *Proceedings of the 20th International Conference on Artificial Intelligence and Statistics*, vol. 54, pp. 1273–1282, Apr. 2017.

[2] T. Taleb, I. Afolabi, K. Samdanis and F. Z. Yousaf, "On multi-domain network slicing orchestration architecture and federated resource control," *IEEE Network*, vol. 33, no. 5, pp. 242–252, Sept.–Oct. 2019, doi: 10.1109/MNET.2018.1800267

[3] A. Thantharate, C. Beard and S. Marupaduga, "A thermal aware approach to enhance 5G device performance and reliability in mmwave networks," *2020 International Symposium on Networks, Computers and Communications (ISNCC)*, 2020, pp. 1–5, doi: 10.1109/ISNCC49221.2020.9297313.

[4] P. Frenger and R. Tano, "More capacity and less power: How 5G NR can reduce network energy consumption," *2019 IEEE 89th Vehicular Technology Conference (VTC2019-Spring)*, Kuala Lumpur, Malaysia, 2019, pp. 1–5.

[5] A. Thantharate, C. Beard and S. Marupaduga, "An approach to optimize device power performance towards energy efficient next generation 5G networks," *2019 IEEE 10th Annual Ubiquitous Computing, Electronics & Mobile Communication Conference (UEMCON)*, 2019, pp. 0749–0754, doi: 10.1109/UEMCON47517.2019.8993067.

[6] The 3rd Generation Partnership Project (3GPP) https://www.3gpp.org (Last accessed 01/11/2023).

[7] A. Thantharate, R. Paropkari, V. Walunj and C. Beard, "DeepSlice: A deep learning approach towards an efficient and reliable network slicing in 5G networks," *2019 IEEE 10th Annual Ubiquitous Computing, Electronics & Mobile Communication Conference (UEMCON)*, New York City, NY, USA, " 2019, pp. 0762–0767.

[8] A. Thantharate, R. Paropkari, V. Walunj, C. Beard and P. Kankariya, "Secure5G: A deep learning framework towards a secure network slicing in 5G and beyond," *2020 10th Annual Computing and Communication Workshop and Conference (CCWC)*, Las Vegas, NV, USA, 2020, pp. 0852–0857.

[9] Y. Liu, X. Yuan, Z. Xiong, J. Kang, X. Wang and D. Niyato, "Federated learning for 6G communications: Challenges, methods, and future directions," *China Communications*, vol. 17, no. 9, pp. 105–118, Sept. 2020.

[10] S. Niknam, H. S. Dhillon and J. H. Reed, "Federated learning for wireless communications: Motivation, opportunities, and challenges," *IEEE Communications Magazine*, vol. 58, no. 6, pp. 46–51, Jun. 2020.

[11] Z. Yu et al., "Federated learning based proactive content caching in edge computing," *2018 IEEE Global Communications Conference (GLOBECOM)*, Abu Dhabi, United Arab Emirates, 2018, pp. 1–6.

[12] C. Zhang, P. Patras and H. Haddadi, "Deep learning in mobile and wireless networking: A survey," *IEEE Communication Surveys and Tutorials*, vol. 21, no. 3, pp. 2224–2287, thirdquarter 2019.

[13] G. Wang, G. Feng, T. Q. S. Quek, S. Qin, R. Wen and W. Tan, "Reconfiguration in network slicing—optimizing the profit and performance," *IEEE Transactions on Network and Service Management*, vol. 16, no. 2, pp. 591–605, June 2019.

[14] N. H. Tran, W. Bao, A. Zomaya, M. N. H. Nguyen and C. S. Hong, "Federated learning over wireless networks: Optimization model design and analysis," *IEEE INFOCOM 2019 - IEEE Conference on Computer Communications*, Paris, France, 2019, pp. 1387–1395.

[15] W. Shi, S. Zhou, Z. Niu, M. Jiang and L. Geng, "Joint device scheduling and resource allocation for latency constrained wireless federated learning," *IEEE Transactions on Wireless Communications*, vol. 20, no. 1, pp. 453–467, Jan. 2021. https://doi.org/10.1109/TWC.2020.3025446

[16] R. Fantacci and B. Picano, "When network slicing meets prospect theory: A service provider revenue maximization framework," *IEEE Transactions on Vehicular Technology*, vol. 69, no. 3, pp. 3179–3189, Mar. 2020.

[17] Q. Xu, J. Wang and K. Wu, "Learning-based dynamic resource provisioning for network slicing with ensured end-to-end performance bound," *IEEE Transactions on Network Science and Engineering*, vol. 7, no. 1, pp. 28–41, 1 Jan.–Mar. 2020.

[18] B. Brik and A. Ksentini, "On predicting service-oriented network slices performances in 5G: A federated learning approach," *2020 IEEE 45th Conference on Local Computer Networks (LCN)*, 2020, pp. 164–171.

[19] S. Yu, X. Chen, Z. Zhou, X. Gong and D. Wu, "When deep reinforcement learning meets federated learning: intelligent multitimescale resource management for multiaccess edge computing in 5G ultradense network," *IEEE Internet of Things Journal*, vol. 8, no. 4, pp. 2238–2251, 15 Feb. 2021.

[20] Y. Zhan, P. Li and S. Guo, "Experience-driven computational resource allocation of federated learning by deep reinforcement learning," *2020 IEEE International Parallel and Distributed Processing Symposium (IPDPS)*, pp. 234–243, 2020. doi: 10.1109/IPDPS47924.2020.00033.

[21] Z. Yang, M. Chen, W. Saad, C. S. Hong and M. Shikh-Bahaei, "Energy efficient federated learning over wireless communication networks," *IEEE Transactions on Wireless Communications*, vol. 20, no. 3, pp. 1935–1949, Mar. 2021. doi: 10.1109/TWC.2020.3037554.

[22] C. T. Dinh et al., "Federated learning over wireless networks: Convergence analysis and resource allocation," *IEEE/ACM Transactions on Networking*, vol. 29, no. 1, pp. 398–409, Feb. 2021. doi: 10.1109/TNET.2020.3035770.

[23] Y. Sun, S. Zhou and D. Gündüz, "Energy-aware analog aggregation for federated learning with redundant data," *ICC 2020 - 2020 IEEE International Conference on Communications (ICC)*, 2020, pp. 1–7. doi: 10.1109/ICC40277.2020.9148853.

[24] M. Usama and M. Erol-Kantarci, "A survey on recent trends and open issues in energy efficiency of 5G", *Sensors*, vol. 19, no. 14, pp. 3126, Jul. 2019.

[25] S. Buzzi, Chih-Lin I., T. E. Klein, H. V. Poor, C. Yang and A. Zappone, "A survey of energy-efficient techniques for 5G networks and challenges ahead", *IEEE Journal on Selected Areas in Communications*, vol. 34, no. 4, pp. 697–709, Apr. 2016.

[26] A. Bohli and R. Bouallegue, "How to meet increased capacities by future green 5G networks: A survey," *IEEE Access*, vol. 7, pp. 42220–42237, 2019, doi: 10.1109/ACCESS.2019.2907284.

[27] A. Thantharate, "FED6G: Federated chameleon learning for network slice management in beyond 5G systems," *2022 IEEE 13th Annual Information Technology, Electronics and Mobile Communication Conference (IEMCON)*, 2022, pp. 0019–0025, doi: 10.1109/IEMCON56893.2022.9946488.

[28] A. Thantharate, V. Walunj, R. Abhishek and R. Paropkari, "Balanced5G - Fair traffic steering technique using data-driven learning in beyond 5G systems," *2022 8th International Conference on Advanced Computing and Communication Systems (ICACCS)*, Coimbatore, India, 2022, pp. 01–07, doi: 10.1109/ICACCS54159.2022.9785169.

[29] GREEN6G Dataset is available at https://www.kaggle.com/datasets/anuragthantharate/green6g-chameleon-federated-learning-for-5g

Chapter 5

Green machine learning approaches for cloud-based communications

Mona Bakri Hassan
Sudan University of Science and Technology (SUST), Khartoum, Sudan

Elmustafa Sayed Ali Ahmed
Sudan University of Science and Technology (SUST), Khartoum, Sudan
Red Sea University (RSU), Port Sudan

Rashid A. Saeed
Taif University, Taif, Saudi Arabia

5.1 INTRODUCTION

Lately, the number of data centers has been increased, and it is well known that data centers consume a high volume of energy. Cloud computing technology is network service that can be provided through internet as data storage and processing facility. It permits consumers to access the processing resource by utilizing a centralization model and infrastructure. Furthermore, it provides computation resources via a virtual environment through the internet [1]. Data center contains several machines based on a company or business firm that offer services and applications including servers, cables, devices, uninterruptable power supply (UPS), air conditions, etc. These machines are heavily working during the day and night in the data centers and consume a large volume of power and energy, then release a high volume of carbon dioxide. The main challenge that may face the Cloud communications (CC) provisioning is the energy utilization and efficiency enhancement. In the past decade, CC energy utilization and efficiency have been addressed extensively in the literature. Therefore, the Green Cloud Computing (GCC) idea was invented recently to discuss the energy consumption issues.

Several algorithms, protocols, and techniques were contributed to decrease the CC energy consumption [2]. ML technique helps to improve cloud computing operations by enhancing power utilization, predicting data flow, and performing control actions based on the running state of cloud computing systems [3].

The methods of optimizing cloud-based energy models for cloud communications depend on the ability to search for different input parameters such as virtual machine (VM) relay, in addition to allocating resources to get the appropriate parameter settings that help reduce power consumption.

DOI: 10.1201/9781003230427-5

In general, one of the most important techniques that reduce energy is the use of clustering methods. In clustering VM-based networks, selecting clusters heads (CHs) is sensitive for analogous nodes tasks to achieve the operations with acceptable energy utilization and consumption [4]. However, these approaches traditionally do not alter the input parameters for CC power consumption. Therefore, ML algorithms are used as a model because of their advantages that help cloud parameters to train the model and forecast the energy consumptions risks. Optimization methods based on ML approaches enable to finding of the appropriate input parameters for clustering-based networks and optimizing the selection CHs to perform tasks, which will help to reduce the high-energy consumption risks.

A few years back, several cloud applications in telecommunications networks have been developed and implemented, which provide services related to digital databases and cloud computing in the field of energy management and other social services through cloud data centers [5]. Therefore, machine and deep learning mechanisms have become amazing techniques that receive great attention from researchers, especially their use as strategies that help improve the data center cloud in terms of power management to achieve green energy in cloud communications.

Green ML approaches for green cloud-based communications topics are growing very fast in industry and academia, telecommunication industries, vendors, and operators [6]. GML becomes one of the key players in the cloud storage and processing of telecom data. This makes energy and power consumption severe challenges in this area. The main aim of this chapter is to discuss the concepts of green ML approaches for cloud-based communications.

The GCC is a quite new topic that has been recently addressed by many researchers. These researches are still scattered, where a critical survey or analysis of the proposed approaches is needed for evaluation. Very few researches were conducted for ML for GCC; most of the research conducted was for ML for CC or GCC without looking for the data processing mechanisms. This chapter aims to conduct a critical evaluation and review related to ML for GCC.

According to the important impact of the ML and DL approaches in cloud-based communications, this chapter addresses the significant ML/DL methods to enable green cloud-based communications energy efficiency. This chapter contribution is focusing on:

- Providing a comprehensive literature review of the existing methods and approaches used to improve energy utilization in cloud-based communications.
- Explain green ML approaches in cloud communications, with emphasis on energy usage and consumption, resources management, and energy efficiency (EE).
- Finally, reviews the future trend GML cloud-based data centers and its challenges.

This chapter is structured as follows: GCC-related works and technical background are discussed in Section 5.2. Energy efficiency of GCC-based ML techniques and solutions is reviewed in Section 5.3, and GCC resource management is presented as well. The reinforcement learning (RL) techniques and methods utilized in green cloud communications are discussed and elaborated in Section 5.4. In Section 5.5, the DL approaches for energy-efficient GCC and cloud computing aspects are reviewed. Section 5.6 presents the challenges and future vision in cloud data centers. Finally, Section 5.7 gives concluding remarks for this chapter.

5.2 GREEN CLOUD COMPUTING BACKGROUND AND RELATED WORKS

CC is one of the hottest and dynamic fields of research and information communication technology (ICT). Its models and methods have various application domains since it offers, trustworthiness, reliability, and scalability, and it provides enhanced performance with reasonable cost. CC architecture is associated with many challenges related to energy management, virtualization, quality of service, security, and standardization. Also, management and energy-aware issues in green cloud communication that have been received high attention in the last decade [4]. The term GCC from uses' perspective can be identified as servers cluster and architecture over the internet whose goals are to decrease the energy data consumption with acceptable performance [5]. The greatest important advantage of GCC is decreasing the consumption of energy. It permits administration, live migration of VMs (virtual machines), and placement optimization. In particular, GCC is a modern and broad field concerned with the mechanisms of energy resource management for cloud computing between consumers and service providers that are concerned with the ecosystem. Usually, services required by users are provided through the cloud in real-time with the possibility of storing information and interoperability protocols for devices and distributed computing in the network. It relates to energy and how it is used. All of the process and task scheduling algorithms, cooling systems, networking, and storage devices are integrated into a framework that connects all operations between these parts at the lowest energy cost. Many researches focus on efficiency in CC energy consumption taking the considerations of task scheduling, storage, cooling, resource allocations, and AI technologies regarding the achievement of green energy requirements.

A detailed discussion of the current literature review that utilizes ML/AI to afford energy efficiency (EE) and new outcomes in CC infrastructures is discussed in [6]. In addition, the chapters shows comparative benchmarking for the proposed solutions for the topic. Furthermore, author provides a perusal for non-ML/AI techniques review for data centers' consumption of energy and the utilization of ML in various CC objectives. Moreover, one of the chapters in [6] proposes methods to reduce the cooling expenses by using distribution of thermal resolution from several load assignments. ML/AI is also employed for energy conservation by optimum CPU frequencies construction.

A framework proposal is presented in [7], which discussed the energy efficiency performance and violation prevention for Services Level Agreement (SLA) by customer for the resources of the implemented green cloud (GC) resources implementation. The framework accomplishes cloud infrastructure resource allocation in Fuzzy Logic and reinforcement learning mechanism for green solutions. The authors evaluated energy efficiency for the allocated resources on the PlanetLab software traces that virtualize the GCC environment for achieving better CPU utilization and Power Usage Effectiveness (PUE). The authors claim that the invented enhanced cloud data center's energy efficiency (EE) utilizes three different parameters, i.e., the efficiency of the infrastructure of data center, PUE, and finally CPU utilization.

A brief concept about using bio-inspired mapping algorithms for resource scheduling-based VM in cloud data centers is disused in [8]. The authors also evaluate the performance of the VMs mapping method considering several parameters such as consumption of energy, QoS, and resource utilization. In addition, they present a novel algorithm for real workflow in the cloud data center by analyzing the bio-inspired approach. Results obtained from new algorithm were gave better efficiency when compared with the other bio-inspired methods.

A centralized resource allocation techniques are introduced by [9] by employing ML that guarantees avoidance of interferences and maximizes EE through preserving QoS requirements for all customers. In addition, the authors consider the user's needs in resource blocks (RBs) allocation and built-in state illustrations based on ML methods to improve the process of learning. Through the analytical results of the resource allocation algorithm used in the study, it was found that it could avoid interferences, and significantly enhance energy and spectrum efficiencies in addition to maintaining the users' requirements for QoS.

A brief definition is viewed in [10], which presented GCC and its development; the overviews and literature cover the GCC integration of architectures, frameworks, innovation, and techniques, in addition to a review of ML applications toward energy-efficient management of CC environment. Moreover, the authors discussed a relative categorization of the proposed algorithms. The authors also surveyed non-ML methods for data center's conservation of energy, and the application of ML/AI toward other procedures in the cloud. Furthermore, one of the chapters in [10] proposes some methods to decrease the cost of the cooling focusing on the decentralize thermal by employing various placements for workload.

The core deliberations of green communication (GC) and the related works on ML/AI-based GC are well reviewed in [11]. The authors focused on ML/AI-based network management and enhancing the harvesting of energy toward GC as well as analyzing the cooperation of state-of-the-art for ML/DL approaches with legacy AI approaches and analytical models for reducing the complexity of the algorithm and accuracy rates optimization for 6G applications acceleration. Lastly, one of the chapters in [11] discussed the present challenges and problematic issues for 6G emergent technologies. Then, an analysis of green economic benefits is given in [12], which is achieved by perusing the ML pertinent data and critical challenges in the CC hypothetical models. In addition, it provided a development reference for theory of economic effectiveness to assist the industry to articulate cities strategies for green growth (Table 5.1).

In addition to what was discussed in the previous studies, there are other requirements related to green cloud computing such as production, operation, and recycling. It is possible to recycle and reuse electronic waste, especially equipment that contains lead, as it helps reduce the impact on the environment. In general, using more efficient and less emitting resources to reduce carbon emissions helps significantly reduce the impact on the environment; in addition to that, pooling resources and reducing the number of devices significantly reduce energy consumption.

Table 5.1 Related works summary

Years	*Approaches*	*Features*	*Advantages*
[9], 2015	Resource allocation centralized techniques by using ML	Resource blocks (RBs) Priorities allocation by using ML	Enhancing EE while preserving QoS
[10], 2015	ML applications in energy-efficient management for CC environment	Present a relative categorization of the proposed approaches	Decreasing cost of cooling by considering non-centralized thermal that produced by various load placement.
[6], 2019	EE-based ML in could computing	The proposed method for cooling costs reduction based on various workload assignments	Optimal construction of CPU frequencies
[7], 2020	The framework of EE and preventing SLA defilement in GCC	Cloud's infrastructure resource allocation-based RL/Fuzzy Logics for green technologies	PUE and CPUs consumptions
[11], 2021	Analyze that ML and DL techniques cooperate with conventional AI methods in 6G applications	Present challenges and problems for 6G emerging technologies.	Accuracy rate optimization for 6G services acceleration and enhance energy harvesting toward the green era.

(Continued)

Table 5.1 (Continued)

Years	*Approaches*	*Features*	*Advantages*
[12], 2021	Analyzes the green economic benefits by discussing the related ML data and other vital challenges	Economic efficiency development and assist industry to articulate strategies for green cities development	Regional and urban overall development.
[8], 2022	Bio-inspired mapping algorithms for resource scheduling in cloud data centers	Novel algorithm-based locust-inspired approach to decrease consumption of energy the number of VMs migration	Outperformance than other bio-inspired algorithms

5.2.1 Green cloud computing (GCC)

The concept of GCC defines an efficient resource utilization in the cloud to minimize the power consumption as much as possible. In data centers, cooling and computation green systems are deployed to efficiently utilize energy [13]. The computation resources approaches are accountable for more than 50% of the power utilization in data centers. An EE in resources management mechanism plays vital role in computing processes and data manipulation and provides an enhanced workload flow [14]. Some of these factors are energy consumptions, traffic workloads, and financial issues. The GCC is providing an enhancement in terms of the cited issues, which can lead to reduce the consumed energy, workload balance, and return of investment (ROI) enlargement [15].

The GCC energy budgets are usually carefully computed based on energy limitations and usage. Particularly, in high workload traffic, ineffective power management has further influence on the energy budget calculation due to the work overload [16]. This will increase the power utilized by the resources [17]. Figure 5.1 shows EE model with other GCC environment factors.

Virtualization is one of the great tools that has been used in GCC recently in the literature for an efficient resource utilization and management [18]. Virtualization is defined as a scheme of the formation of virtual resource to be used by data center's cloud [19], which can improve hardware machines efficiently and provide efficient resources sharing. One of the most utilized virtualization tools is the VMs that manage the resource efficiently, enable remote interface, monitor, and control the virtualization processes with cloud security management [20, 21].

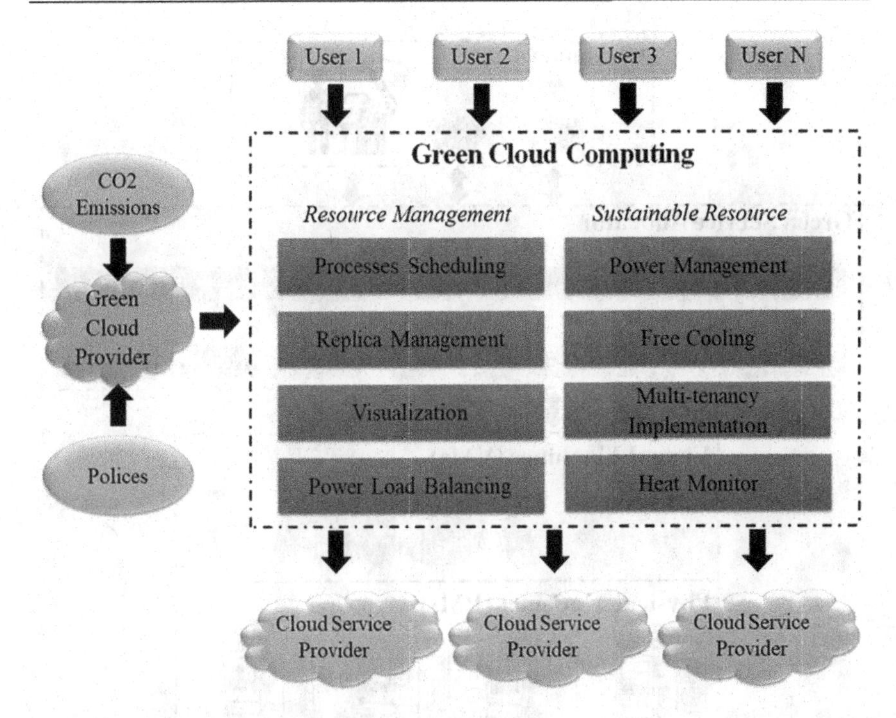

Figure 5.1 Green cloud computation architecture.

In the GCC, various communication services can be provided by cloud requests through other middleware known as a green authority, which enables the provision of cloud services related to programming, and infrastructure with high-quality guarantees [22]. As shown in Figure 5.2, the cloud infrastructure consists of different cloud services that enable access to customer information from afar while providing green cloud computing to enable the management of complete applications in addition to managing energy in servers, storage, and VM as well [23].

Moving on to the concept of GCC, many features can be gained such as the ability to expand, balancing load between the intra and inter clouds, in addition to high reliability [24, 25]. The GCC methodologies related to energy efficiency and cloud computing sustainability are related to processing, scheduling, and computation virtualization.

A. **Computation Processing**
   Microprocessors play an important role in cloud computing operations regarding power consumption versus processing, as the technologies of the processor industries may significantly reduce power consumption levels [25]. Companies are currently competing in the production of processors with very high speeds, as new technologies

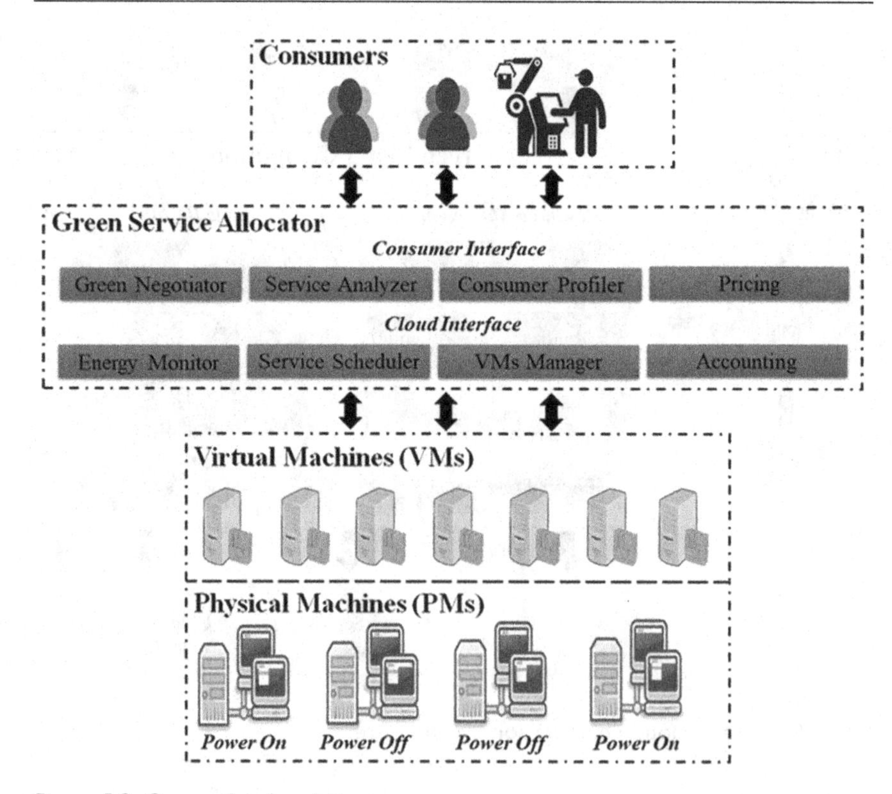

Figure 5.2 Green cloud architecture.

and industrial materials are used to reduce energy consumption during data processing.

B. **Computation Task Scheduling**
Scheduling operations and tasks greatly helps in reducing energy consumption in cloud computing. Computational fluid dynamics models are one of the most important methods that help in scheduling tasks versus thermal modeling, where thermal scheduling is given for data center tasks [23, 25]. Thermal scheduling algorithms control the temperature within the data center through task assignment, reducing cooling requirements.

C. **Computation Virtualization**
Computer resource virtualization helps create multiple virtual machines (VMs) on a physical server, effectively reducing the hardware's use of computing resources that consume more power. This method greatly improves the maintenance of private computing infrastructure with the help of online power management mechanisms that control the effects of various power management policies by these virtual machines on the default resource [23, 25].

5.2.2 Machine learning for cloud communications

Cloud is used for computing, networking, and storage. The integration of cloud and ML algorithms will increase the benefits of using the cloud significantly [26]. In general, ML algorithms take time to implement some tasks, but with the cloud computing model, ML tasks can be processed more quickly. Cloud computing has become more popular due to the efforts of famous companies such as Amazon, Microsoft, and Google, they allow a large amount of data to increase exponentially that will leads to increased opportunities for utilization [27]. There are several advantages of using ML in cloud computing as shown in the following:

A. The cloud pay-per-use model is a useful solution for companies that used AI or ML workloads without spending a ton of money.
B. The cloud makes it possible for the enterprise to test and measure ML capabilities as projects go into production as well as increase demand.
C. The cloud provides access to mutual abilities without the need for advanced skills in AI or data science.

There are some applications based on ML that use a cloud computing approach such as cognitive computing, personal assistance, business intelligence (BI), and AI-as-a-Service (AIaaS). In cognitive computing, a large amount of data is stored in the cloud. These data are used as a source for ML algorithms. Companies used the cloud for storage, data sharing, and networking [27]. But recently, ML algorithms have been used to learn from data and achieve results in a shorter time. These applications can provide cognitive processes and expect outcomes. Using AI/ML in cognitive computing will find applications in healthcare, marketing, banking, and finance, and many fields [28].

For personal assistance, chatbots have become popular as a personal assistant and customer support service, i.e., Microsoft Cortana, Google Assistant, and Apple's Siri. The chatbots are built using NLP and ML technologies. Personal assistants have limitations in their capabilities. When chatbots and personal assistants collaborate with the cloud, it will lead to access to large amounts of data [29]. With the cloud's mass data, ML's ability to learn from data and its cognitive computing feature, personal can transform any user interaction. These skills will be great to run companies and businesses [30]. The system can find all the information about past transactions, analyse current sales, and make expectations for profits in the future. In addition, it explains if the function went wrong and what can be done to fix it.

Business intelligence services will become smarter by applying ML in cloud computing. ML and cloud computing help BI enterprises to achieve their goals by processing and analysing real-time data, and predicting for future. It allows you to create an interactive dashboard that displays data from various dimensions in one place [31]. The integration of ML algorithms with cloud computing will give a positive effect on the current state of BI services. Machine Learning Business Intelligence will help to achieve that goal.

For AI-as-a-Service (AIaaS) applications, AI enables the provision of intelligent applications for cloud providers through open-source platforms. It provides a group of AI tools for the functionality that users need. AIaaS is a delivery model that supports effective cost for AI solutions rather than consulting many AI professionals to finish the tasks [32]. Also, it is a service that simplifies the process of intelligent automation for humans so that they do not need to get into trouble with functionality that will enhance the capabilities of cloud computing and increase the requirement for using AI with the cloud.

5.2.3 Cloud communication (CC) energy utilization

Recently, due to the increasing use of cloud communications (CC), energy consumption has become large, which requires a method to reduce its consumption in CC, as it has become one of the hot research topics among academic and industrial researchers [33]. The studies prove that CC consumes about 3% of global consumption for energy which results in high operation cost for cloud service providers that leads to an increase in the total ownership cost for infrastructures of data centers [34]. Also, the carbon dioxide (CO_2) emission is one of great issues have been raised recently that is associated with CC operation and new research findings for the CO_2 emission [35]. Figure 5.3 shows the power consumption volume in the data center will rise continuously [36].

Extensive research has been conducted in CC to reduce power consumptions volume by hardware machine resources in data centers and hence the overall CC system enhancement in terms of performance. In CC, the machines continue to provide web services [37] which consume high power to perform efficiently. Some of the research has used the term GCC to reduce the power consumptions in CC [38, 39]. Also, physical machines in data centers could produce hazardous heat and gases, in which GCC is introduced to deal with such issues and for e-waste management. The power consumption volume can be measured by two methods: hardware and/or software [40].

5.3 ML FOR GCC

Cloud computing can provide several services, i.e., SaaS i.e., Slack, Microsoft Office 365, Gmail, and PaaS, IaaS, FaaS (Function as a Services) [41], LooS (logging as a Services), i.e., web-services of Amazon, Azure of Microsoft, and compute-engine of Google [42]; CaaS (Container as a Services), DaaS (Data

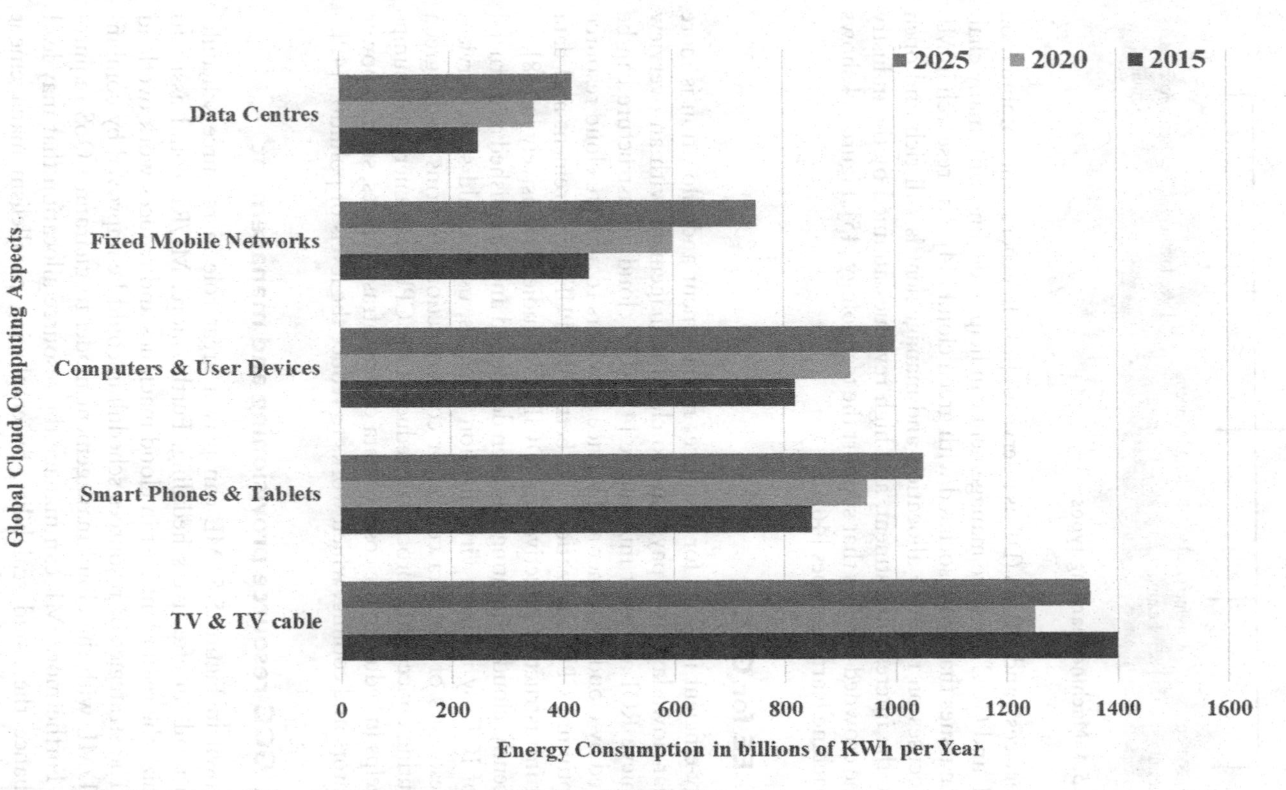

Figure 5.3 Forecast energy consumption of global cloud computing.

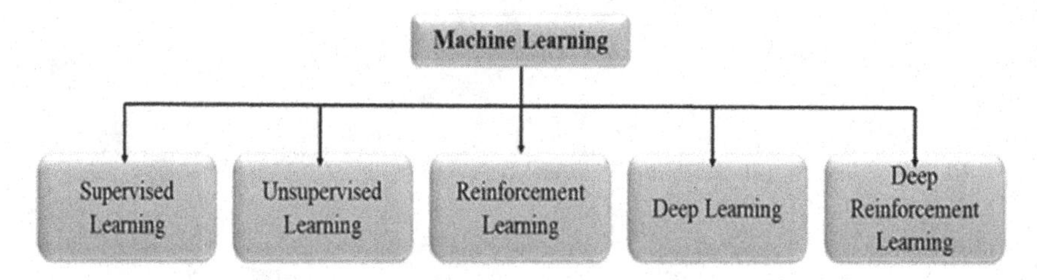

Figure 5.4 Machine learning types.

as a services), and DBaaS (DB as a services) [43]. In general, the combination of ML and GCC resource management can help in dealing with many challenging issues that are associated with green cloud [44]. The research in ML that focuses on resource allocation and management is still rich and open due to the interest, investment, and high revenues attained by the industry and the crowded services that support the technology [45]. Figure 5.4 shows the machine learning types [46].

5.3.1 EE for GCC

Energy-efficient in GCC for resources management and allocation is more popular nowadays as it pays care to cloud management with an overview to achieve ROI and cost minimalize [47]. The cloud infrastructure can be inspired by cloud customer and service providers to allocate cloud resource for optimum CPUs utilization and better EE. PUE measurement and data center infrastructure effectiveness (DCiE) are studied extensively in [48].

Recently, cloud computing has been developed and established as a crucial part of ICT. By utilizing virtualization schemes, users could share services and resources of the data centre for cost reduction. In terms of the green computing concept, the process of reducing the CPU peak energy consumption helps in reducing the costs of data centre infrastructures, such as power generators and cooling, which works to reduce the levels of pollution [49].

5.3.2 GCC resource provisioning and management

As shown in Figure 5.5, ML can play a great role in resource schedule. It can result in effective scheduling. Furthermore, ML/RL could assist to minimize the redundant cost in cloud machines and reduces work overload [50]. The strategies of resources scheduling could be improved by combining RL/ML with the cloud management model to guarantee QoS requirement. Furthermore, ML can manage the resource allocation that may lead to enhance the cloud scalability, by employing parallelism management approach to minimize the computing complexity. In addition, workload predictor based on RL/ML algorithm can enhance the actions processes of the cloud green servers [51].

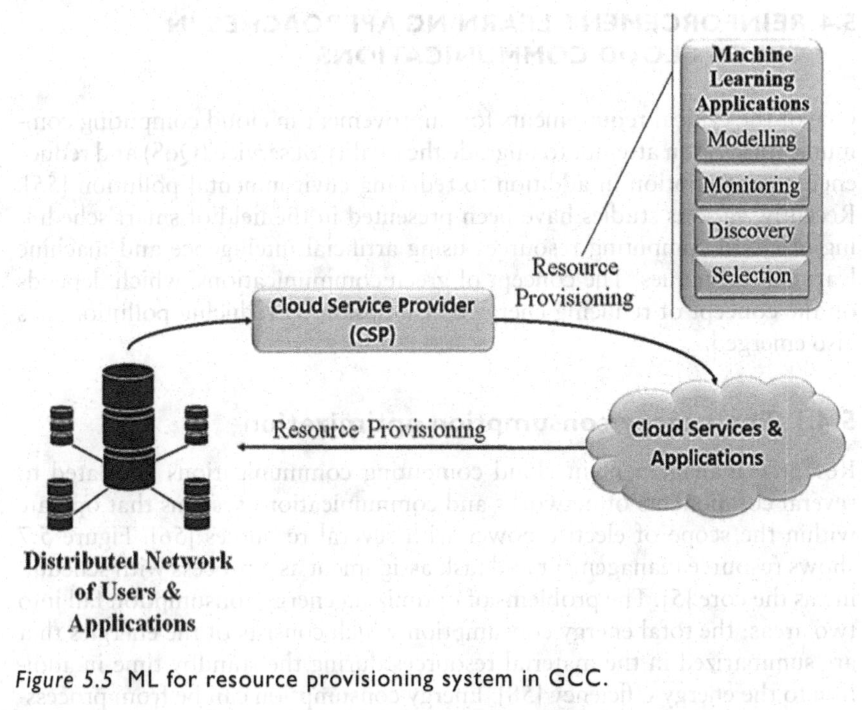

Figure 5.5 ML for resource provisioning system in GCC.

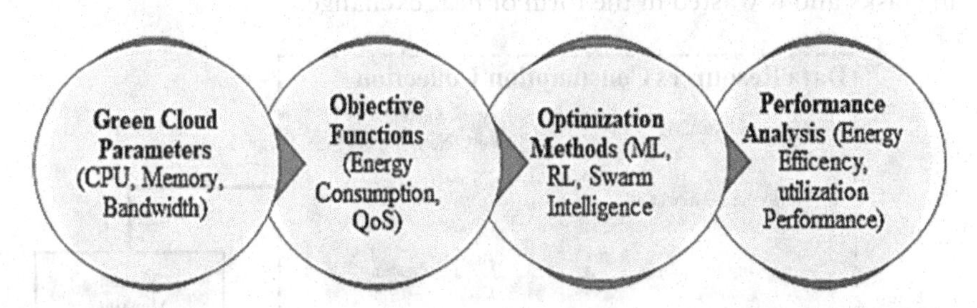

Figure 5.6 The EE optimization process in cloud resources management scheme.

Recently, several researches have appeared concentrating on cloud resources management concept for GCC. They deal with the likelihood of utilizing DL, RL, and ML methods to build a smart computing system with high EE in resources management [52]. In GCC resources management, EE is the key concern and issue in data centers [53]. The green schedule-based DL/RL/ML schemes will assist to forecast service level and then conserve energy up to 50% or more. Figure 5.6 presents the DL, RL and ML concept by utilizing an optimization technique for green cloud EE. Such schemes and approaches are applied to optimize and to reduce the energy consumptions and enhance the QoSs [54].

5.4 REINFORCEMENT LEARNING APPROACHES IN GREEN CLOUD COMMUNICATIONS

One of the critical requirements for improvement in cloud computing communications is an attempt to upgrade the quality of service (QoS) and reduce energy consumption in addition to reducing environmental pollution [55]. Recently, various studies have been presented in the field of smart scheduling of cloud computing resources using artificial intelligence and machine learning techniques. The concept of green communications, which depends on the concept of reducing energy consumption by reducing pollution, has also emerged.

5.4.1 The energy consumption optimization

Resource management in cloud computing communications is related to several components of networks and communications systems that operate within the scope of electric power with several resources [56]. Figure 5.7 shows resource management and task assignment as a process with scheduling as the core [5]. The problems of optimizing energy consumption fall into two areas: the total energy consumption which consists of the energies that are summarized in the material resources during the standby time in addition to the energy efficiency [58]. Energy consumption can be from processing tasks and is wasted in the form of heat exchange.

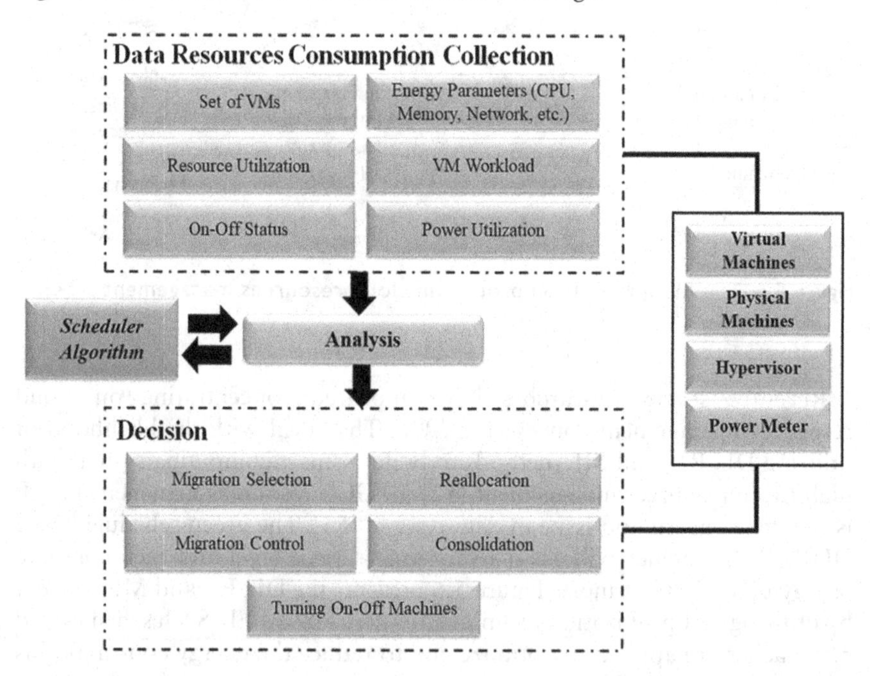

Figure 5.7 The platform of optimization energy consumption.

Recently, many researchers have been made for intelligent cloud communication-based 5G technology applications. AI provides an improvement to the wireless terminals and networks as well. The enhancement of AI in communication networks is achieved by enabling automatic and adaptive resource management processes [59]. RL enables to optimization of the network decisions based on the data collected from the environment. In 5G cloud-based communications, RL provides a means of intelligent control adaptively and automatic management. The RL algorithms in 5G cloud communications provide an enhancement in three application levels: mobile broadband, low latency, and massive communications [60]. For these applications, the requirement is to achieve a reliable and adaptive network to ensure many services of high quality.

In 5G cloud-based communications, mobile edge computing (MEC), with network virtualization impact to provide low latency and flexible network services [61]. MEC enables to provide high computation capabilities at the 5G network edge close to the mobile terminals and plays a vital role to act as an adaptive relay to directly make responses to the terminals without core network intersection. This approach will reduce the communications distance to give lower latency and enhance the traffic load [62]. RL promises to enhance the computation capability and enables terminals to adjust the offload operations at the network edge, which will help to reduce the computation delay and improves the quality of experience (QoE).

For the caching computation, many policies are used to enable caching hit rate. Such policies like least recently used (LRU) enable the improvement of the replacement scheme. However, with massive data and long-term caching, the commutation requires to be optimized based on the prior knowledge of data [63]. In addition, caching adjustment is required in real-time network requests such in IoT-based 5G networks, to enhance the learning capability of cache policy. Due to the learning approach, cache computing takes place by making the system decide to compute the data depending on the state of the system resources. The use of RL in caching enables to predict the future mobile terminal requests and hence, the average reward of terminal demands can be maximized accordingly. The cache policies based on RL can be applied to the huge data in multicast MEC [64]. RL approach enables the provision of the best knowledge of future terminal requests and improves the cache management mechanism with high accuracy.

5.4.2 Reinforcement learning for 5G cloud-based architecture

In 5G IoT-based cloud computing, the quality of user experience is very important in both aspects, latency and power consumption. The limitation of computing capabilities will affect the QoS in case of high traffic loads. Therefore, an intelligent task offloading procedure is important in such a situation to efficiently compute the workload of IoT terminals. During cloud, computing, mobile terminals can control the offload until the task is finished [65]. This strategy will reduce the edge nodes' power consumption during cloud computations. The use of RL in 5G IoT-based cloud architecture will improve the edge computing close to the terminals to provide short communications with enhanced tasks offloading scheme. Moreover, the RL approach in the offloading mechanism will enable the mobile terminal to adjust offloading to prevent attacks [66]. The mobile terminal can be possibly interfered with by attackers and they must carefully perform the suitable offloading rate to be safe.

In industrial IoT (IIoT)-based 5G networks, the extension of networking is increased due to the attractive technology, and hence, many devices, wireless sensors, control models, and remote terminal units are used to improve industrial efficiency [67]. In such an application, the IIoT network environment will be facing different issues related to connectivity, spectrum resources, and communication delay. The 5G IIoT cloud-based network architecture, as shown in Figure 5.8, creates high communication latency between cloud and industrial devices which requires solutions to be overcome [68]. In such architecture, edge computing can reduce network congestion, and enable resource optimization. In addition, edge computing is easy to be deployed with IIoT in the case of massive devices. The use of DL algorithms promises to achieve great enhancement in large-scale 5G IIoT networks [69]. DL helps to observe the data processing and enables high accuracy decision making.

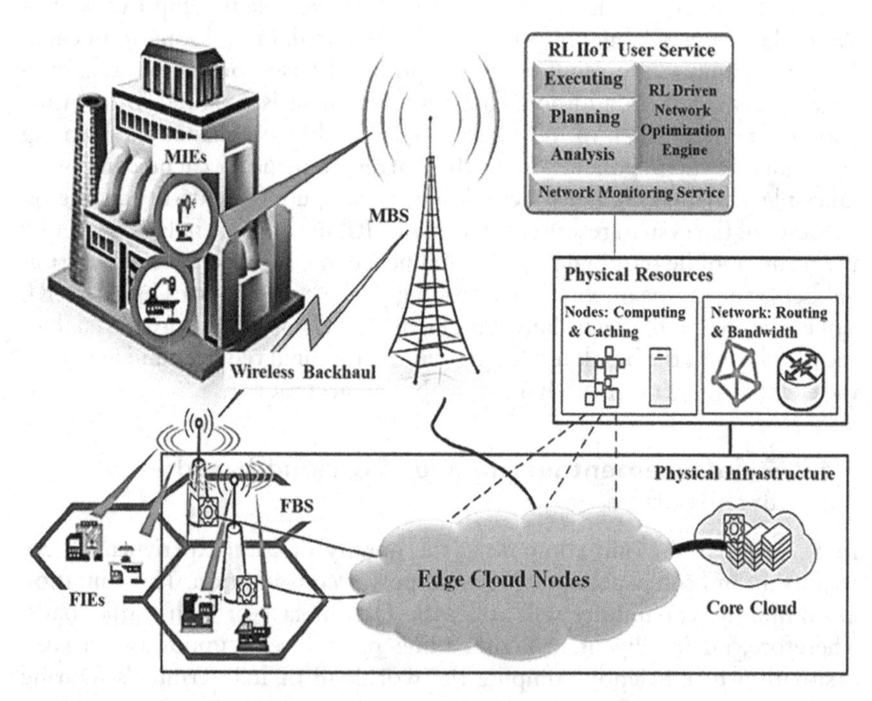

Figure 5.8 RL in green 5G cloud-based architecture for IIoT networks.

5.4.3 Reinforcement learning for scheduling decisions

The architecture of the green 5G cloud based on DL provides intelligent features extraction of industrial data and predicts multiple metrics by training a large quantity of data, which will remain to successfully learn the cloud and edge scheduling policies according to traffic flow [70]. RL can provide an efficient workload scheduling approach by learning from previous actions and optimizing the scheduling process. In the IIoT, RL enables the state of the system environment to intelligently let the multi-tier edge computing system to selects accurate parameters for the system environment to make scheduling decisions. For optimization in action rewards expectations, the Q-Learning algorithm enables the two adjustments of the Q-values as a function for rewards to select the best action according to the Q-values [71]. The framework of RL containing action selection and strategy update is shown in Figure 5.9.

In vehicular cloud communication, the roadside units (RSUs), and vehicles embedded unit; a.k.a. in-board units (IBU), are the key machines in the

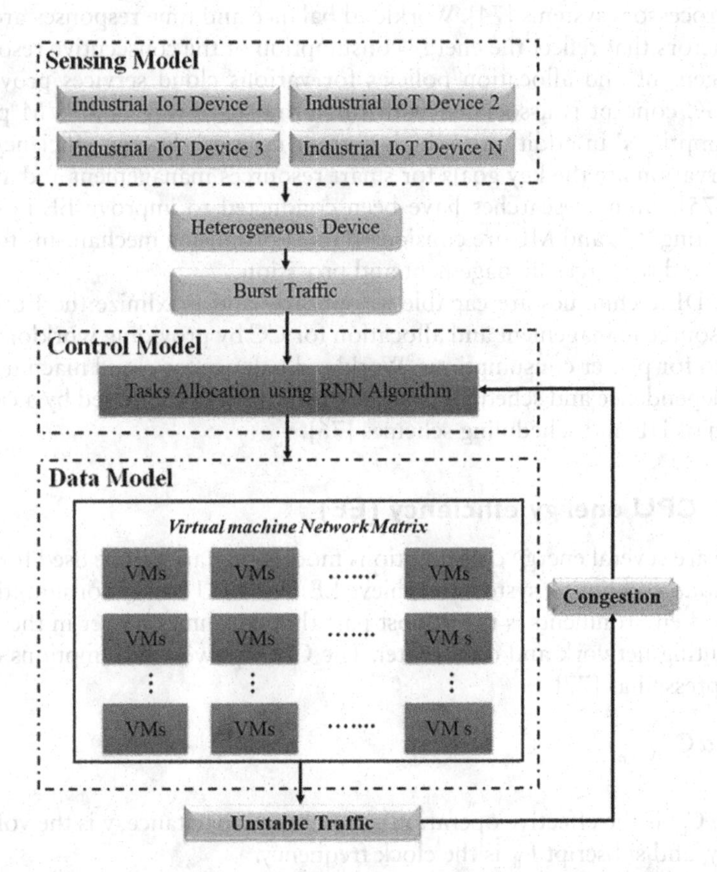

Figure 5.9 Reinforcement learning framework for scheduling decisions in IIoT.

network and required cache processing and resources management and allocation. These machines are related rapidly used and update by the rapidly moving vehicles, which leads to many challenges in resource allocation and QoS requirements [72]. These problematic issues are usually influenced by traffic loads, service SLA, and traffic classes. The usage of AI/RL/ML mechanisms may help for dynamic resource provisioning and allocation by providing smart scheduling decision-making [73]. The usage VM in the vehicular network/cloud also help to improve the resource assignment. Ai/RL/ML for VMs allocation may maximize and optimize the resources employment in the long term.

5.5 DEEP LEARNING APPROACHES FOR ENERGY EFFICIENT IN GREEN CLOUD COMPUTING

In cloud network, the resource management and allocation performance will significantly depend on a few factors, i.e., processing energy of the hardware machines associated with the cloud, i.e., storage machines, buses, peripheral and processors systems [74]. Workload balance and time responses are crucial factors that reflect the energy consumption of the consecutive resources management and allocation policies for various cloud services providers. The GC concept is associated with the terms 'hardware' and 'VM power consumptions' in addition to 'high current utilities'. Energy efficiency and conservation are the key goals for smart resources management and allocation [75]. Many researches have been conducted to improve EE in cloud computing. DL and ML are considered most promising mechanisms to provide cloud resources management and provision.

ML/DL techniques are capable to optimize and maximize the EE index for resource management and allocation for CC by providing workload balance to for power consumptions. Workload balance for cloud machines is a task dependence and scheduling issue which could be improved by a combination of DL and scheduling schemes [76].

5.5.1 CPU energy efficiency (EE)

There are several energy consumptions models that are can be used for CPU and cloud computing system to achieve EE. The CPU energy consumption in physical environments is the utmost part that consumes power in the cloud computing network and data center. The CPUs power consumptions could be expressed as [77]:

$$\mathrm{E}\,\alpha\,\mathrm{C}_{\mathrm{effV}^2\mathrm{f}_{\mathrm{clk}}} \tag{5.1}$$

where $\mathrm{C}_{\mathrm{eff}}$ is the effective operation's switching capacitance, v is the voltages supply, and subscript $\mathrm{f}_{\mathrm{clk}}$ is the clock frequency.

The utilization of CPU after allocating loads to the VMs can be expressed as [78]:

$$E(u) = (P_{max} - P_{min}) \times \frac{u}{100} + P_{min} \quad (5.2)$$

where E(u) is the energy consumption and it's stated as CPU utilization, u is the instance microprocessor utilizations amount, the P_{min} and P_{max} are the minimum and maximum energy consumptions in watts, respectively.

Energy inefficiencies in cloud-based data centers are related to the power consumption that is wasted when running servers with weak CPUs and can cause low data reliability, in addition to poor performance even with low loads [76, 78]. Whereas weak CPUs can consume more energy, almost exceeding 50% of the given power to avoid the performance bottleneck problem [76]. In terms of the green computing concept, the process of reducing the CPU peak energy consumption helps in reducing the costs of data centre infrastructures, such as power generators and cooling, which works to reduce the levels of pollution.

The improvement of CPU power efficiency and applications of power-saving techniques enables the operation of the green cloud computing-based data centre in low efficient power mode and dominates the server's energy consumption. The correlation between the CPU utilizations and energy consumptions by a server expressed by Equation 5.2 shows that the energy consumptions by servers is increasing linearly with the increase in power consumption by CPUs [78]. The enabling of good dynamic power management (DPM) policies helps to forecast the actual volume of energy consumptions based on some of the runtime system characteristics.

5.5.2 Resource allocation and management

Resource allocation and management have been well discussed and extensively studied in the ICT arena, for efficient resources utilization and cost-effective purposes. VMs are an emulation of computer systems and considered as one of the cloud resources, which need to be allocated and assigned for users' and consumers' tasks [79]. The tasks originated from users and customers that need to be allocated and assigned to VMs rely on number of parameters, i.e., available resource, CPU usage, access time, etc.

Resource allocation and management depend on scheduling mechanism, which applies resource schedule based on certain criteria and policies such as task priority or task energy consumption. The allocation schemes are utilized to enable optimization for cloud workload distribution with high performance and an efficient execution [80]. Figure 5.10 shows the resources scheduling model/framework, which presents the cloud security infrastructure (CSI) that performs an authentication and authorization processes for

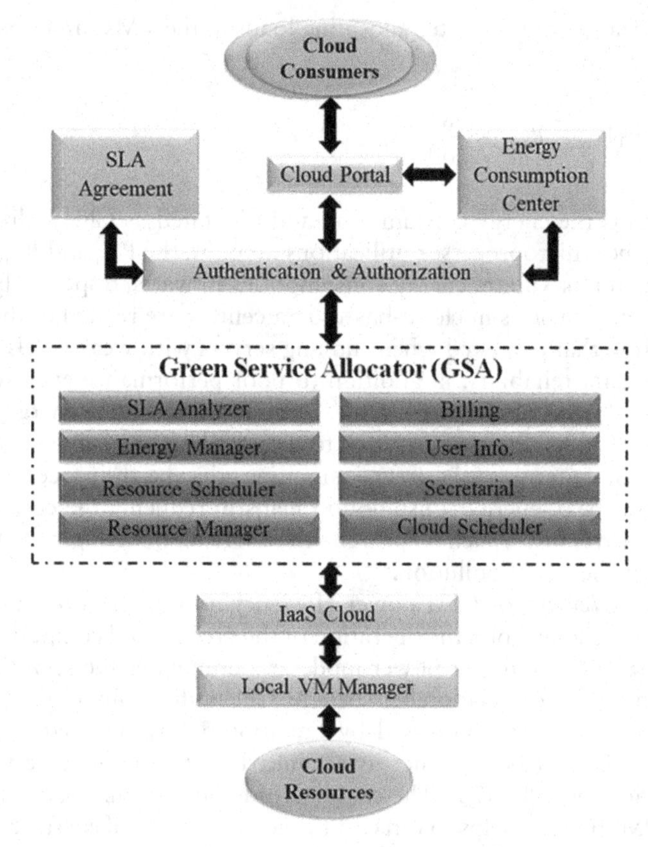

Figure 5.10 Energy-efficient resource scheduling framework.

cloud customers to be able to access the cloud resources and presents the energy consumptions, QoS parameters, and service levels agreement (SLAs).

The green services allocations (GSAs) work collaboratively with applications and services among local resources scheduling and then enabling green resource allocation. The scheduling algorithm data is manipulated by the resource allocation manager and then processed by the workload schedulers of the cloud [81]. The resource scheduler is a logical entity that acquires the cloud users' requirements and adjusts the workload scheduler decisions. The availability of resources is tracked by the resource's allocation manager, and its response to relocating resources across physical machines will improve the service allocation judgments.

Various scheduling mechanisms have been newly invented and enhanced to provide resource allocation, management, and control in data centre and cloud [82]. The policies and procedures have been improved as well for EE in cloud resource scheduling and assignment to provide high grade of services and SLA to customers with low power consumption. Other researches are focused on dynamic allocation for VM with minimum power consumption. Furthermore, studies were conducted and proved that a combination of SLA resources provisioning predictively and reactively can lead to energy consumption reduction by more than 36% [83]. This methodology results of long-term analysis for workload demand by manage resource allocator and scheduler to satisfy the SLA requirement.

5.5.3 Energy consumption in green cloud computing

In the cloud computing and data centers, the most aspects manipulating energy consumptions are the energy distributions, cooling systems, hardware and soft computing traffic load distribution. All these aspects need to be well considered, studied, and optimized to ensure that optimized energy efficiency has been adopted [84]. The usage of intelligent and smart approaches like ML/DL/AI algorithms greatly assist to overcome many of the issues related to resources management and allocation in cloud environment. These techniques are widely applied in various services and application for optimization and prediction of resources sharing and allocation. They have been utilized extensively in cloud computing literature as new trends of cloud-based optimized services. The cloud also categorized and split to edge and fog computing for efficient resources utilization and for optimistic service provisioning [85]. The cloud load prediction helps to pause and sleep the unutilized machines and then energy conservation that can result from intelligent approaches to provide optimum decisions for future required scheduling.

The massive data with an adaptive environment such as a data centre is hard to work optimally for many reasons as follows.

- Equipment is heating and power consumption is following a non-linear function.
- The centers usually are not designed to adapt to external or internal variations. This needs unlimited scenarios to be covered.
- The centers have various environments and architecture which leads to difficulties to design a unique solution for each architecture.

To study these issues, recently ML and DL approaches have been applied to the data centre and cloud computing more competently. For example, DeepMind® and Google's data centers develop a system that enhanced the center's utilities dramatically [85]. The team developed a trained NN for various operating situations and factors. With the neural networks, the system worked more efficiently and adaptively to adapt, accommodate, and optimize the dynamicity of the system efficiently.

A deep neural network (DNN) can do many complex cognitive tasks, enhancing the performance dramatically compared to classical ML algorithms [86], whereby they often need a massive volume of data to train and could be very computationally intensive. By using deep learning on cloud computing, a large amount of data is allowed to be easily managed and ingested to train algorithms as well as it enables deep learning models to scale efficiently and at the lowest costs using GPU processing power [87]. Cloud computing services help to make deep learning more allowable as well as it is easier to manage huge data volume and train algorithms on distributed hardware. Cloud services provide deep learning in four respects, they are;

- Allow large-scale computing capacity on-demand and make it possible to distribute model training via multiple machines.
- Allow access to special hardware configurations.
- Do not need an upfront investment.
- Assist deep learning management workflows.

Recently, DL technique advancements have enabled DNNs to be implemented in various applications field. DL has approved great results in speech recognition, computer vision, and natural language processing [88]. DeepMind Technologies is a British company for artificial general intelligence. DeepMind created a NN for video games to imitate humans, as well as a neural turning machine or a NN that can access an external machine as a legacy/traditional turning machine, toward a computer that can imitate the human memory [89]. Many projects handled by DeepMind® as:

5.6 CLOUD DATA CENTERS' FUTURE TRENDS AND CHALLENGES

Today, most of the provided services to bank customers, students, national civil information, utility services, etc. are cloud computing–based. These services are co-operating with each other and share databases and other resources among themselves efficiently which becomes and considered as one of the national wealth that need to be well managed and protected [92]. Some of the typical cloud computing examples include Microsoft Online, Google App Engine, YouTube, and Apple iCloud others. Services as online banking, shopping, email, and social media are most popular cloud computing applications. However, this massive deployment of cloud computing and data centers comes with many issues and challenges that need to be addressed and studied to solve issues such as power consumption, security, management, and cost [93].

5.6.1 Cloud computing energy's efficiency problems

Discussions of energy is an important issue in cloud computing system due to the consumed energy volume and expenses that are paid to cloud computing that has massively been deployed all over the globe. The researchers invented the term GCC to overcome and manipulate the problems and challenges of energy in the cloud [94]. GCC is the procedures of modeling, engineering, manufacturing, and allocating of data centers resources with minimum computing expenses. Some enterprises and giant companies took serious actions to adopt friendly and efficient cloud computing models that have the green capabilities and by using renewable power sources, transfer the servers' heat to other types of energy to be used for other services, and to ensure that all data centre and cloud hardware portions are healthy working and to be recycled when they no longer work efficiently and consuming more power [95].

Various countries are investigating GCC and reviewing the energy consumption effect to telecom electricity system and data centers. Concerning the ICT industry studies in India and China and India, there will be high growth rate for cloud power consumptions with about 63% by the end of 2025 [96]. As presented in Figure 5.11, the ICT industry showed statistics that the overall cloud energy consumed between 2008 and 2021 reaches its highest levels. The statistic provides energy consumed in terms of carbon dioxide metric tons (MtCO2) equivalent that are released to the environment for each cloud object [97]. The overall cloud power consumptions in 2030 are anticipated to be at vital levels. Concerning the presented statistics, in the near future the EE should be applied to all services and application in the cloud to overcome the energy effect into environment and guarantee and certify the GCC abilities.

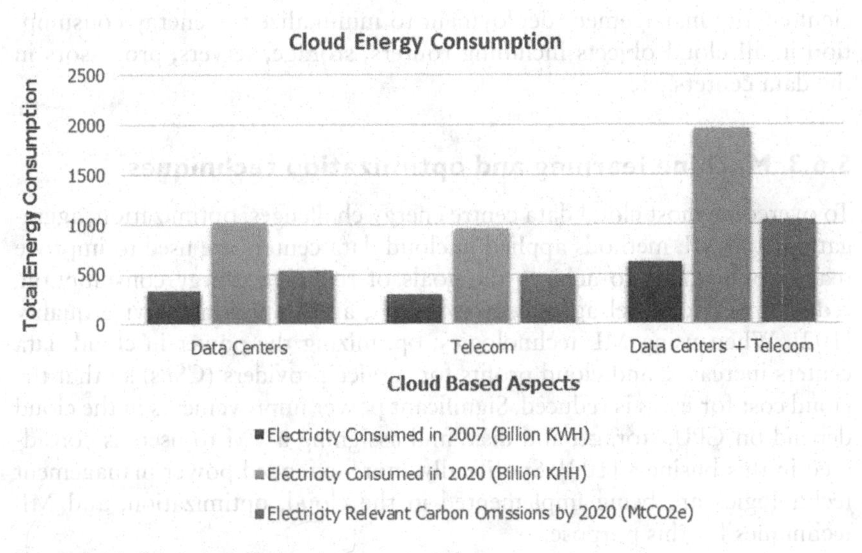

Figure 5.11 Statistics of cloud energy total consumptions.

The decentralized data centre using cloud computing consumes a large volume of energy. There are two communal approaches to decrease the energy volume in cloud computing system: by balancing servers' overload and pausing the servers in data centers [97–98]. There are numerous approaches and methods that are utilized to resolve the energy inadequacy problems such as voltages scaling, dynamic frequencies, VM consolidation and migrations, high GCC grow, energy expansion maximum bin-pack and min-max, and migrations minimization [99]. The main aim of all these approaches/algorithms is EE maximization in clouds. Dynamic energy management (DEM) system could be utilized to mitigate massive energy unnecessary usage in cloud computing. This algorithm at the beginning uses only the essential cloud resources, and then estimates the power amount needed and provisions the suitable power voltages [100]. If any cloud resources are provisioned with the required energy, the DPM algorithm will correct and optimize the adequate power, a.k.a. dynamic energy optimization (DEO) procedure.

5.6.2 Cloud computing virtualization

In the idea of GCC, consolidation is defined as the process deploying several data centers associated to data processing services on a single server with virtualization technologies [101]. The consolidation is considered as sub-task, and it is realized by using load balancing at process level, enhanced virtual system utilization as well as decreasing energy consumption. Energy efficiency (EE) is an essential GCC's building block which contributes a key role at the eco-system green cloud architecture [102]. Energy efficiency (EE) in cloud communications could be denoted as efficient energy management deployment to minimalize the energy consumption in all cloud objects including routers, storage, servers, processors in the data centers, etc.

5.6.3 Machine learning and optimization techniques

To overcome most cloud data centre energy challenges, optimization, aggregation, and ML methods applied in cloud data centers are used to improve many applications to achieve the goals of reducing energy consumption, reducing service level agreement violation, and improving service quality [102]. When using ML technologies, optimizing the power in cloud data centers increases, and cloud profits for service providers (CSPs) so that the cloud cost for users is reduced. Significant power improvements in the cloud depend on CPU, storage, and disk, and assigning a VM to users is considered in this business [103]. Specifically, more extended power management technologies are being implemented in the cloud, optimization, and ML techniques for this purpose.

5.7 CONCLUSION

Many researches presented the recent studies and contributions in the areas of GCC, GC, and ML and the advantages that could be gained and attained from the services related to them. Among these techniques are the utilization of cloud- and edge computing-based processing and the utilization of ML to structure advanced GCC systems. This chapter has primarily deliberated many current state-of-the-art smart, green communication based on CC with the opportunity of employing DL/ML methods such as training and mechanisms techniques that assist to build vastly effective systems in terms of energy consumptions and performance. This chapter also summarized the emerging DL technique performance in the initial stages and deep potential results to address the communications challenges based on GCC in addition to a suggestion of existing opportunities, bottlenecks, and future visions.

Through this chapter, the reader can learn the general aspects related to secure communication systems on green smart computing from the fifth generation and know the efficiency of training when applying learning algorithms to them. Although deep learning is not mature in wireless communications, it is a powerful tool and a hot research topic in many potential areas of future application. Processes such as channel estimation, data analysis and mobility, complex decision making, network management, and resource optimization all fall within the benefits of using deep learning and developing analytics and applications for advanced cloud computing.

REFERENCES

[1] Mehmet Demirci, "A survey of Machine Learning Applications for Energy-Efficient Resource Management in Cloud Computing Environments." *International Conference on Machine Learning and Applications (IEEE)*, 2015.

[2] Zeinab E. Ahmed, Rashid A. Saeed, Amitava Mukherjee, "Vehicular Cloud Computing models, architectures, applications, challenges, and opportunities." *Ch 3*, in *Vehicular Cloud Computing for Traffic Management and Systems*, IGI Global, USA, pp. 57–74, June 2018.

[3] Mona B. Hassan, E. S. Ali, R. A. Saeed, *Chapter 5: Intelligence IoT Wireless Networks; Book: Intelligent Wireless Communications*, 2021, Chap. 6, pp. 135–162, IET book publisher.

[4] Mohamed Deiab, Deena El-Menshawy, Salma El-Abd, Ahmad Mostafa, M. Samir Abou El-Seoud, "Energy Efficiency in Cloud Computing." *International Journal of Machine Learning and Computing*, Vol. 9, No. 1, February 2019.

[5] I. AlQerm, B. Shihada, "Enhanced Machine Learning Scheme for Energy Efficient Resource Allocation in 5G Heterogeneous Cloud Radio Access Networks." *IEEE 28th Annual International Symposium on Personal, Indoor, and Mobile Radio Communications (PIMRC)*, 2017, pp. 1–7.

[6] Pratibha Pandey and Abhishek Singh, "Energy Efficient Resource Management Techniques in Cloud Environment for Web-Based Community by Machine

Learning: A Survey." *International Journal of Web Based Communities*, Vol. 15 No.3, 2019.

[7] Thandar Theina, Myint Myat Myoa, Sazia Parvinb, Amjad Gawanmeh, "Reinforcement Learning Based Methodology for Energy-Efficient Resource Allocation in Cloud Data Centers," *Journal of King Saud University – Computer and Information Sciences*, 2020.

[8] Ala'anzy, M.A., Othman, M. "Mapping and Consolidation of VMs Using Locust-Inspired Algorithms for Green Cloud Computing." *Neural Processing Letters*, vol. 54, pp. 405–421, 2022.

[9] Mahmoud Al-Ayyoub, Mohammad Wardat, Yaser Jararweh, Abdallah A. Khreishah, "Optimizing expansion strategies for ultra-scale cloud computing data centers." *Simulation Modelling Practice and Theory*, Vol. 58, No. Part 1, pp. 15–29, 2015.

[10] Mehmet Demirci, "A survey of Machine Learning Applications for Energy-Efficient Resource Management in Cloud Computing Environments." *International Conference on Machine Learning and Applications (IEEE)*, 2015.

[11] Bomin Mao, Fengxiao Tang, Kawamoto Yuichi, Nei Kato, "AI based Service Management for 6G Green Communications." *Networking and Internet Architecture*, 2021.

[12] Zhang Jin, "*Retracted Article: Green City Economic Efficiency Based on Cloud Computing and Machine Learning*." Springer Link, 2021.

[13] Azad, P., & Navimipour, N. J. "An Energy-Aware Task Scheduling in the Cloud Computing using a Hybrid Cultural And Ant Colony Optimization Algorithm." *International Journal of Cloud Applications and Computing*, Vol. 7, No. 4, pp. 20–40, 2017.

[14] Shuja, J., Ahmad, R.W., Gani, A. et al., "Greening Emerging IT Technologies: Techniques and Practices." *Journal of Internet Services and Applications*, Vol. 8, p. 9, 2017. https://doi.org/10.1186/s13174-017-0060-5

[15] Sara A. Mahboub, Ali ES Ahmed, RA Saeed, "Smart IDS and IPS for Cyber-Physical Systems." In *Artificial Intelligence Paradigms for Smart Cyber-Physical Systems*, edited by Luhach, Ashish Kumar, Atilla Elçi, pp. 109–136. Hershey, PA: IGI Global, 2021.

[16] A. Mijuskovic, A. Chiumento, R. Bemthuis, A. Aldea, P. Havinga, "Resource Management Techniques for Cloud/Fog and Edge Computing: An Evaluation Framework and Classification." *Sensors (Basel)*, Vol. 21, No. 5, pp. 1832, 2021.

[17] R. S. Abdalla, S. A. Mahbub, R.A. Mokhtar, E.S. Ali, R.A. Saeed, "IoE Design Principles and Architecture." *Internet of Energy for Smart Cities: Machine Learning Models and Techniques*. CRC Press Publisher, 2020. https://doi.org/10.4018/978-1-7998-5101-1

[18] Nimisha Patel, Hiren Patel, "Energy Efficient Strategy for Placement of Virtual Machines Selected from Underloaded Servers in Compute Cloud." *Journal of King Saud University - Computer and Information Sciences*, Vol. 32, No. 6, pp. 700–708, 2020.

[19] Elmustafa S. Ali Ahmed, Rasha E.A. Elatif, "Network Denial of Service Threat Security on Cloud Computing -A Survey." *International Journal of Scientific Research in Science, Engineering and Technology (IJSRSET)*, Vol. 1, No. 5, 2015.

[20] Xialin Liu, Junsheng Wu, Gang Sha, Shuqin Liu, "Virtual Machine Consolidation with Minimization of Migration Thrashing for Cloud Data Centers." *Mathematical Problems in Engineering*, Vol. 2020, Article ID 7848232, 2020.

[21] M.B. Hassan, E. S. Ali, N. Nurelmadina, R.A. Saeed, "Artificial Intelligence in IoT and its Applications' (Telecommunications, 2021), 'Intelligent Wireless Communications." Chap. 2, pp. 33–58, IET Digital Library.
[22] R. Boutaba, M. A. Salahuddin, N. Limam et al., "A Comprehensive Survey on Machine Learning for Networking: Evolution, Applications and Research Opportunities." *Journal of Internet Services and Applications*, Vol. 9, p. 16, 2018. https://doi.org/10.1186/s13174-018-0087-2
[23] U. A. Butt, M. Mehmood, S. B. H. Shah, R. Amin, M. W. Shaukat et al., "A Review of Machine Learning Algorithms for Cloud Computing Security." *Electronics*, 9, 1379, 2020.
[24] I. H. Sarker, Machine Learning: Algorithms, "Real-World Applications and Research Directions". *SN Computer Science*, Vol. 2, p. 160, 2021.
[25] Jacqueline Schmitt, Jochen Bönig, Thorbjörn Borggräfe, Gunter Beitinger, Jochen Deuse, "Predictive model-based quality inspection using Machine Learning and Edge Cloud Computing." *Advanced Engineering Informatics*, Vol. 45, 2020.
[26] Nahla Nurelmadina, Mohammad Kamrul Hasan, R. A. Imran Memon Khairul Akram Saeed et al., "A Systematic Review on Cognitive Radio in Low Power Wide Area Network for Industrial IoT Applications," *Sustainability*, MDPI, Vol. 13, No. 1, pp. 1–20, 2021.
[27] K. K. Nirala, N. K. Singh, V. S. Purani, "A Survey on Providing Customer and Public Administration Based Services using AI: Chatbot." *Multimedia Tools and Applications*, 2022.
[28] Zeinab E. Ahmed and R. A. Saeed, Amitava Mukherjee, "Challenges and Opportunities in Vehicular Cloud Computing." In *Cloud Security: Concepts, Methodologies, Tools, and Applications*, edited by Management Association, Information Resources, pp. 2168–2185. Hershey, PA: IGI Global, 2019.
[29] Sururah A. Bello, Lukumon O. Oyedele, Olugbenga O. Akinade, Muhammad Bilal, Juan Manuel Davila Delgado et al., "Cloud Computing in Construction Industry: Use Cases, Benefits and Challenges." *Automation in Construction*, Vol. 122, p. 103441, 2021.
[30] Lina E. Alatabani, E. S. Ali, R.A. Mokhtar, R. A. Saeed, Hesham Alhumyani, M. K. Hasan, "Deep and Reinforcement Learning Technologies on Internet of Vehicle (IoV) Applications: Current Issues and Future Trends." *Journal of Advanced Transportation*, Vol. 2022, Article ID 1947886, 2022.
[31] Muhammad Rizwan Anawar, Shangguang Wang, Muhammad Azam Zia, Ahmer Khan Jadoon, Umair Akram, Salman Raza, "Fog Computing: An Overview of Big IoT Data Analytics." *Wireless Communications and Mobile Computing*, Vol. 2018, Article ID 7157192, 2018.
[32] N. M. Elfatih, M. K. Hasan, Z. Kamal, D. Gupta, R. A. Saeed, E. S. Ali, M. S. Hosain, "Internet of Vehicle's Resource Management in 5G Networks using AI Technologies: Current Status and Trends." *IET Communications*, pp. 1–21, 2021. https://doi.org/10.1049/cmu2.12315
[33] M. Z. Ahmed, A. H. A. Hashim, O. O. Khalifa, R. A. Alsaqour, RA Saeed, A. H. Alkali, "Queuing Theory Approach for NDN Mobile Producer's Rate of Transmission Using Network Coding," *2021 International Congress of Advanced Technology and Engineering (ICOTEN)*, pp. 1–6,2021.
[34] A. S. G. Andrae, T. Edler, "On Global Electricity Usage of Communication Technology: Trends to 2030." *Challenges*, Vol. 6, pp. 117–157, 2015. https://doi.org/10.3390/challe6010117

[35] Fahad A. Alqurashi, F. A. Alsolami, S. Abdel-Khalek, Sayed Ali Elmustafa, R.A. Saeed, “Machine Learning Techniques in Internet of UAVs for Smart Cities Applications'. *Journal of Intelligent Fuzzy Systems*, pp. 1–24, 2021, https://doi.org/10.3233/JIFS-211009

[36] Monali Jumde, Snehlata Dongre, “Analysis on Energy Efficient Green Cloud Computing.” *International Conference on Research Frontiers in Sciences*, 2021.

[37] Anton Beloglazov, Rajkumar Buyya, Young Choon Lee, Albert Zomaya, “Chapter 3 - A Taxonomy and Survey of Energy-Efficient Data Centers and Cloud Computing Systems.” *Advances in Computers*, Vol. 82, pp. 47–111, 2011.

[38] Michael Maurer, Ivona Brandic, Rizos Sakellariou, “Adaptive Resource Configuration for Cloud Infrastructure Management.” *Future Generation Computer Systems*, Vol. 29, No. 2, pp. 472–487, 2013.

[39] Faroug M. Osman Nagi, and Ali Ahmed A. Elamin, Ali E. S. Ahmed, R. A. Saeed, “Cyber-Physical System for Smart Grid.” In *Artificial Intelligence Paradigms for Smart Cyber-Physical Systems*, IGI Global, 2021. https://doi.org/10.4018/978-1-7998-5101-1.ch014

[40] Gustavo Caiza, Morelva Saeteros, William Oñate, Marcelo V. Garcia, “Fog Computing at Industrial Level, Architecture, Latency, Energy, and Security: A Review.” *Heliyon*, Vol. 6, No. 4, p. e03706, 2020.

[41] Shah SC. “Design of a Machine Learning-Based Intelligent Middleware Platform for a Heterogeneous Private Edge Cloud System.” *Sensors (Basel)*, 2021.

[42] S. K. U. Zaman, A. I. Jehangiri, T. Maqsood et al., “Mobility-Aware Computational Offloading in Mobile Edge Networks: A Survey.” *Cluster Computing*, Vol. 24, pp. 2735–2756, 2021.

[43] R. A. Abbas Alnazir Hesham Alhumyani Mokhtar, E. S. Ali, R. A. Saeed, S. Abdel-Khalek. “Quality of Services Based on Intelligent IoT WLAN MAC Protocol Dynamic Real-Time Applications in Smart Cities.” *Computational Intelligence and Neuroscience*, Vol. 2021, Article ID 2287531, 2021.

[44] Jianyu Wang, Jianli Pan, Flavio Esposito, Prasad Calyam, Zhicheng Yang, Prasant Mohapatra, “Edge Cloud Offloading Algorithms: Issues, Methods, and Perspectives.” *ACM Computing Surveys*, 52, 1–23, 2019.

[45] Yu Wei et al., “A Survey on the Edge Computing for the Internet of Things.” *Special Section on Mobile Edge Computing*, Vol. 6, 2018.

[46] M. Jayachandran, C. Rami Reddy, Sanjeevikumar Padmanaban, A. H. Milyani, “Operational Planning Steps in Smart Electric Power Delivery System.” *Scientific Reports*, Vol. 11, p. 17250, 2021. https://doi.org/10.1038/s41598-021-96769-8

[47] Shahab Shamshirband et al., “Game Theory and Evolutionary Optimization Approaches Applied to Resource Allocation Problems in Computing Environments: A Survey.” *Mathematical Biosciences and Engineering*, Vol. 18, No. 6), pp. 9190–9232, 2021.

[48] D. Sara, M. Hicham, “Green Cloud Computing: Efficient Energy-Aware and Dynamic Resources Management in Data Centers.” *International Journal of Advanced Computer Science and Applications*, Vol. 9, No. 7, 2018.

[49] Chenn-Jung Huang, Yu-Wu Wang, Chih-Tai Guan, Heng-Ming Chen, Jui-Jiun Jian, “Applications of Machine Learning to Resource Management in Cloud Computing.” *International Journal of Modeling and Optimization*, Vol. 3, No. 2, April 2013.

[50] Ahmadreza Montazerolghaem, Mohammad Hossein Yaghmaee, Alberto Leon-Garcia. "Green Cloud Multimedia Networking: NFV/SDN based Energy-efficient Resource Allocation." *IEEE Transactions on Green Communications and Networking*, IEEE, 2020, pp. 1–1.
[51] Mona Bakri Hassan and Ali ES Ahmed, R. A. Saeed, "Machine Learning for Industrial IoT Systems." In *Handbook of Research on Innovations and Applications of AI, IoT, and Cognitive Technologies*. IGI Global, 2021. https://doi.org/10.4018/978-1-7998-6870-5.ch023
[52] Goyal, Shanky et al., "An Optimized Framework for Energy-Resource Allocation in a Cloud Environment based on the Whale Optimization Algorithm." *Sensors (Basel, Switzerland)*, Vol. 21, No. 5, p. 1583, 24 Feb. 2021, doi:10.3390/s21051583
[53] D. Yang, H. R. Karimi, O. Kaynak, S. Yin. "Developments of Digital Twin Technologies in Industrial, Smart City and Healthcare Sectors: A Survey." *Complex Engineering Systems*, Vol. 1, p. 3, 2021. http://dx.doi.org/10.20517/ces.2021.06
[54] Nawaf O. Alsrehin, Ahmad F. Klaib, Aws Magableh, "Intelligent Transportation and Control Systems Using Data Mining and Machine Learning Techniques: A Comprehensive Study." *IEEE Access*, Vol. 7, 2019.
[55] I. Nurcahyani, J. W. Lee. "Role of Machine Learning in Resource Allocation Strategy over Vehicular Networks: A Survey." *Sensors (Basel)*, Vol. 21, No. 19, p. 6542, 30 Sep 2021. https://doi.org/10.3390/s21196542
[56] F. Farahnakian, P. Liljeberg, J. Plosila, "Energy-Efficient Virtual Machines Consolidation in Cloud Data Centers Using Reinforcement Learning," *22nd Euromicro International Conference on Parallel, Distributed, and Network-Based Processing*, 2014, pp. 500–507. https://doi.org/10.1109/PDP.2014.109
[57] Yaohua Sun, Mugen Peng, Yangcheng Zhou, Yuzhe Huang, Shiwen Mao, "Application of Machine Learning in Wireless Networks: Key Techniques and Open Issues." *IEEE Communication Surveys and Tutorials*, 2019.
[58] Muhammad Sohaib Ajmal, Zeshan Iqbal, Farrukh Zeeshan Khan, Muneer Ahmad, Iftikhar Ahmad, Brij B. Gupta, Hybrid Ant Genetic Algorithm for Efficient Task Scheduling in Cloud Data Centers, *Computers and Electrical Engineering*, Vol. 95, p. 107419, 2021.
[59] Rafael Moreno-Vozmediano, Rubén S. Montero, Eduardo Huedo and Ignacio M. Llorente, "Efficient Resource Provisioning for Elastic Cloud Services Based on Machine Learning Techniques." *Journal of Cloud Computing: Advances, Systems and Applications*, 2019.
[60] Guangyao Zhoue et al, "Deep Reinforcement Learning-based Methods for Resource Scheduling in Cloud Computing: A Review and Future Directions." arXiv:2105.04086v1, 2021.
[61] P. Varga, J. Peto, A. Franko et al., "5G support for Industrial IoT Applications - Challenges, Solutions, and Research gaps." *Sensors (Basel)*, Vol. 20, No. 3, p. 828, 2020.
[62] Chu, W., Wuniri, Q., Du, X. et al. "Cloud Control System Architectures, Technologies and Applications on Intelligent and Connected Vehicles: A Review." *Chinese Journal of Mechanical Engineering*, Vol. 34, p. 139, 2021. https://doi.org/10.1186/s10033-021-00638-4
[63] F. A. M. Bargarai, A. M. Abdulazeez, V. M. Tiryaki, D. Q. Zeebaree, "Management of Wireless Communication Systems Using Artificial

Intelligence-Based Software Defined Radio." *International Journal of Interactive Mobile Technologies (iJIM)*, Vol. 14, No. 13, pp. 107–133, 2020.
[64] Mohammed N. Mahdi, Abdul R. Ahmad, Qais S. Qassim, Hayder Natiq, Mohammed A. Subhi, Moamin Mahmoud, "From 5G to 6G Technology: Meets Energy, Internet-of-Things and Machine Learning: A Survey." *Applied Sciences*, Vol. 11, No. 17, p. 8117, 2021. https://doi.org/10.3390/app11178117
[65] Zhonghua Zhang, Xifei Song, Lei Liu, Jie Yin, Yu Wang, Dapeng Lan, "Recent Advances in Blockchain and Artificial Intelligence Integration: Feasibility Analysis, Research Issues, Applications, Challenges, and Future Work." *Security and Communication Networks*, Vol. 2021, Article ID 9991535, 15 pages, 2021.
[66] Fu Yu, Sen Wang, Cheng-Xiang Wang, Xuemin Hong, Stephen McLaughlin, "Artificial Intelligence to Manage Network Traffic of 5G Wireless Networks." *AI For Network Traffic Control*, IEEE Network, November/December 2018.
[67] Y. Meng, M. A. Naeem, A. O. Almagrabi, R. Ali, H. S. Kim "Advancing the State of the Fog Computing to Enable 5G Network Technologies." *Sensors*, Vol. 20, p. 1754, 2020. https://doi.org/10.3390/s20061754
[68] Ichen Qian, Jun Wu, Rui Wang, Fusheng Zhu, Wei Zhang, "survey on Reinforcement Learning Applications in Communication Networks." *Journal of Communications and Information Networks*, Vol. 4, No. 2, Jun. 2019.
[69] Pinjia Zhang, Yang Wu, and Hongdong Zhu, "Open Ecosystem for Future Industrial Internet of Things (IIoT): Architecture and Application." *CSEE Journal of Power and Energy Systems*, Vol. 6, No. 1, March 2020.
[70] F. Liang, W. Yu, X. Liu, D. Griffith, N. Golmie, "Toward Edge-Based Deep Learning in Industrial Internet of Things." *IEEE Internet of Things Journal*, Vol. 7, No. 5, pp. 4329–4341, May 2020. https://doi.org/10.1109/JIOT.2019.2963635
[71] Xianwei Li, Baoliu Ye, "Latency-Aware Computation Offloading for 5G Networks in Edge Computing." *Security and Communication Networks*, Vol. 2021, Article ID 8800234, 15 pages, 2021. https://doi.org/10.1155/2021/8800234
[72] Zeinab El-Rewini, Karthikeyan Sadatsharan, Daisy Flora Selvaraj, Siby Jose Plathottam, Prakash Ranganathan, "Cybersecurity Challenges in Vehicular Communications." *Vehicular Communications*, Vol. 23, p. 100214, 2020.
[73] Da Nurcahyani, Jeong Woo Lee, "Role of Machine Learning in Resource Allocation Strategy over Vehicular Networks: A Survey." *Sensors*, Vol. 21, p. 6542, 2021.
[74] L. E. Alatabani, E. S. Ali, R. A. Saeed "Deep Learning Approaches for IoV Applications and Services. *Intelligent Technologies for Internet of Vehicles. Internet of Things Technology, Communications, and Computing*, 2021. Springer, Cham.
[75] Rodolfo Meneguette, Robson De Grande, Jo Ueyama, Geraldo P. Rocha Filho, and Edmundo Madeira, "Vehicular Edge Computing: Architecture, Resource Management, Security, and Challenges." *ACM Computing Surveys*, Vol. 55, No. 1, Article 4, 46 pages, 2023.
[76] Sergio Pérez, Patricia Arroba, José M. Moya, "Energy-conscious optimization of Edge Computing through Deep Reinforcement Learning and two-phase immersion cooling." *Future Generation Computer Systems*, Vol. 125, pp. 891–907, 2021.

[77] Ali ES and RA Saeed, "Big Data Distributed Systems Management." In *Networking for Big Data Book, Networking for Big Data*, edited by Shui Yu, Xiaodong Lin, Jelena Misic, Xuemin (Sherman) Shen, Chapter 4, pp. 57–71, April 2015.
[78] Adam Kozakiewicz, Andrzej Lis, "Energy Efficiency in Cloud Computing: Exploring the Intellectual Structure of the Research Field and Its Research Fronts with Direct Citation Analysis." *Energies*, Vol. 14, p. 7036, 2021. https://doi.org/10.3390/en14217036
[79] R. Nasim, E. Zola, A. J. Kassler, "Robust Optimization for Energy-Efficient Virtual Machine Consolidation in Modern Datacenters." *Cluster Computing*, Vol. 21, pp. 1681–1709, 2018. https://doi.org/10.1007/s10586-018-2718-6
[80] Shanchen Pang, Kexiang Xu, Shudong Wang, Min Wang, Shuyu Wang, "Energy-Saving Virtual Machine Placement Method for User Experience in Cloud Environment." *Mathematical Problems in Engineering*, Vol. 2020, Article ID 4784191, 9 pages, 2020.
[81] Miyuru Dayarathna, Yonggang Wen, and Rui Fan, "Data Center Energy Consumption Modeling: A Survey." *IEEE Communication Surveys and Tutorials*, Vol. 18, No. 1, First Quarter 2016.
[82] E. S. Ali, R. A. Saeed, "Intelligent Underwater Wireless Communications." In *Intelligent Wireless Communications*, Chap. 11, pp. 271–305, IET Digital Library.
[83] Mukherjee, Amitava, RA Saeed, Sudip Dutta, Mrinal K. Naskar, "Tracking Framework for Software-Defined Networking (SDN)." *Resource Allocation in Next-Generation Broadband Wireless Access Networks IGI Global*, 2017.
[84] Rehenuma Tasnim Rodoshi, Taewoon Kim, Wooyeol Choi, "Resource Management in Cloud Radio Access Network: Conventional and New Approaches." *Sensors*, Vol. 20, p. 2708, 2020. https://doi.org/10.3390/s20092708
[85] Rania Salih Ahmed, Ali ES Ahmed, R. A. Saeed, "Machine Learning in Cyber-Physical Systems in Industry 4.0." In *Artificial Intelligence Paradigms for Smart Cyber-Physical Systems*, edited by Luhach, Ashish Kumar, Atilla Elçi, pp. 20–41. Hershey, PA: IGI Global, 2021. https://doi.org/10.4018/978-1-7998-5101-1.ch002
[86] Shanky Goyal et al, "An Optimized Framework for Energy-Resource Allocation in a Cloud Environment based on the Whale Optimization Algorithm." *Sensors*, Vol. 21, p. 1583, 2021. https://doi.org/10.3390/s21051583
[87] S. Amudha, M. Murali "Deep Learning Based Energy Efficient Novel Scheduling Algorithms for Body-Fog-Cloud in Smart Hospital." *Journal of Ambient Intelligence and Humanized Computing*, Vol. 12, pp. 7441–7460, 2021.
[88] R. A. Mokhtar and R. A. Saeed, "Conservation of Mobile Data and Usability Constraints." In *Cyber Security Standards, Practices and Industrial Applications: Systems and Methodologies, Ch 03*, edited by Z. Junaid, M. Athar, pp. 40–55, USA: IGI Global, ISBN13: 978-1-60960-851-4, 2011.
[89] Christos L. Stergiou, Konstantinos E. Psannis, Brij B. Gupta, "Flexible Big Data Management Through a Federated Cloud System." *ACM Transactions on Internet Technology*, Vol. 22, No. 2, Article 46, 22 pages, May 2022.
[90] I. H. Sarker, "Deep Learning: 'A Comprehensive Overview on Techniques, Taxonomy, Applications and Research Directions'." *SN Computer Science*, Vol. 2, p. 420, 2021. https://doi.org/10.1007/s42979-021-00815-1

[91] Adil Akasha, R. A. Ali Saeed, Elmustafa Sayed, "Smart Interoperability Public Safety Wireless Network." In *Intelligent Wireless Communications*, Chap. 16, pp. 399–418, IET Digital Library. https://doi.org/10.1049/PBTE094E_ch16
[92] K. Sai Manoj, "Major Issues and Challenges in Energy Efficient Cloud Storage Service." *Psychilgy and Education*, Vol. 57, No. 9, 2020.
[93] Elmustafa Sayed Ali Ahmed, Zahraa Tagelsir Mohammed, M. B. Hassan, R. A. Saeed, "Algorithms Optimization for Intelligent IoV Applications." In *Handbook of Research on Innovations and Applications of AI, IoT, and Cognitive Technologies*, edited by Zhao, Jingyuan, V. Vinoth Kumar, pp. 1–25. Hershey, PA: IGI Global, 2021.
[94] E. S. Ali, M. B. Hassan, R. A. Saeed, "Machine Learning Technologies on Internet of Vehicles." In *Intelligent Technologies for Internet of Vehicles. Internet of Things (Technology, Communications, and Computing)*. Springer, 2021.
[95] R. H. Aswathy, P. Suresh, M. Y. Sikkandar, S. Abdel-Khalek, H. Alhumyani, R. A. Saeed et al., "Optimized Tuned Deep Learning Model for Chronic Kidney Disease Classification: CMC-Computers." *Materials & Continua*, Vol. 70, No. 2, pp. 2097–2111, 2022.
[96] M. B. Hassan, E. S. Ali, R. A. Mokhtar, R. A. Saeed, Bharat S. Chaudhari, "NB-IoT: Concepts, Applications, and Deployment Challenges." In *LPWAN Technologies for IoT and M2M Applications*, editors Bharat S. Chaudhari, Marco Zennaro, Chapter 6, Elsevier. ISBN: 9780128188804, March 2020.
[97] Aasheesh Raizada, Kishan Pal Singh, Mohammad Sajid, "Worldwide Energy Consumption of Hyperscale Data Centers: A Survey." *International Research Journal on Advanced Science Hub*, Vol. 2, Special Issue ICAET 11S, pp. 8–15, 2020. https://doi.org/10.47392/irjash.2020.226
[98] Romany F. Mansour, Nada M. Alfar, Sayed Abdel-Khalek, Maha Abdelhaq, RA Saeed, Raed Alsaqour, "Optimal Deep Learning Based Fusion Model for Biomedical Image Classification." *Expert Systems*, June 2021. https://doi.org/10.1111/exsy.12764
[99] M. Z. Ahmed, A. H. A. Hashim, O. O. Khalifa, RA Saeed, R. A. Alsaqour and A. H. Alkali, "Connectivity Framework for Rendezvous and Mobile Producer Nodes Using NDN Interest Flooding," *International Congress of Advanced Technology and Engineering (ICOTEN)*, 2021, pp. 1–5.
[100] O. O. Khalifa, A. A. Omar, M. Z. Ahmed, R. A. Saeed, A. H. A. Hashim, A. N. Esgiar, "An Automatic Facial Age Progression Estimation System," *2021 International Congress of Advanced Technology and Engineering (ICOTEN)*, 2021, pp. 1–6.
[101] C. Rupa, Divya Midhunchakkaravarthy, Mohammad Kamrul Hasan, Hesham Alhumyani, R. A. Saeed. "Industry 5.0: Ethereum blockchain technology based DApp smart contract." *Mathematical Biosciences and Engineering*, Vol. 18, No. 5, pp. 7010–7027, 2021. https://doi.org/10.3934/mbe.2021349
[102] T. K. Rodrigues, K. Suto, H. Nishiyama, J. Liu, N. Kato, "Machine Learning Meets Computation and Communication Control in Evolving Edge and Cloud: Challenges and Future Perspective." *IEEE Communication Surveys and Tutorials*, Vol. 22, No. 1, pp. 38–67, Firstquarter 2020. https://doi.org/10.1109/COMST.2019.2943405
[103] Z. Chang, S. Liu, X. Xiong, Z. Cai, G. Tu, "A Survey of Recent Advances in Edge-Computing-Powered Artificial Intelligence of Things," *IEEE Internet of Things Journal*, Vol. 8, No. 18, pp. 13849–13875, 15 Sept 2021. https://doi.org/10.1109/JIOT.2021.3088875

Chapter 6

Green machine learning for Internet-of-Things

Current solutions and future challenges

Hajar Moudoud
Université de Sherbrooke, Sherbrooke, QC, Canada
University of Technology of Troyes, Troyes, France

Zoubeir Mlika and Soumaya Cherkaoui
Université de Sherbrooke, Sherbrooke, QC, Canada

Lyes Khoukhi
Normandie University, Caen, France

6.1 INTRODUCTION

The Internet-of-Things (IoT) is an interconnected and distributed network of devices or objects that exchange information through wired or wireless communications. These connected objects participate in a plethora of IoT applications and services that have significant innovation potential andbenefits in all spheres of life and the economy. However, IoT objects are already producing a tremendous amount of data that challenges the capabilities of traditional analytics techniques to mine or process it. To overcome these issues, researchers introduced green IoT. Green IoT is a concept that aims to reduce the energy consumption emitted by materials or software. This means that IoT connected devices/objects use an energy-efficient process with the aim of reducing energy consumption, greenhouse effects and minimizing CO_2 emissions. Machine learning (ML) solutions have come to bring intelligence to IoT applications and services and help provide useful information and improve decision-making.

Artificial intelligence (AI) in general and ML in particular have gained considerable momentum over the past few decades due to advances in various fields such as cloud computing, and applications such as object recognition, and gaming. Most IoT systems are increasingly complex, diverse, and heterogeneous; therefore, managing and exploiting these systems to their full potential is a challenge. Several studies have demonstrated the great benefits of using ML to support IoT [1, 2]. Indeed, ML techniques have been

DOI: 10.1201/9781003230427-6

used in various tasks such as regression, classification, and clustering, to provide useful solutions to infer valuable knowledge from the data generated by the IoT. Similarly, ML can be leveraged to help IoT systems achieve better performance.

Despite the advances and benefits that ML has brought to the IoT, these have come at the cost of significant use of computing resources and energy [3]. This is because ML involves repeatedly training models on large datasets, which requires complex mathematical operations to be performed on each piece of data received. The datasets on which ML models are trained are getting larger and larger, resulting in a significant increase in the energy consumed by the training process. According to recent research, the number of computing resources used to train ML models has increased by a factor of 300,000 over the past six years [4]. With the increase in computational energy demand for many ML algorithms, ML systems have the potential to significantly increase carbon emissions. Recent work has demonstrated the potential impact of model training on the carbon footprint. For example, training a machine translation model could generate about the same amount of CO_2 emissions as five cars (626,000 tons of CO_2) [5].

This trend is driven by the AI community, which seeks to attain accurate models that can yield systematic results. Despite the obvious benefit of improving model accuracy, the focus on this one metric does not consider the ecological or environmental cost of achieving the desired accuracy. Many recent ML research neglects to present the energy and carbon footprint of model training. We refer to these environmentally unfriendly ML systems as "red ML." Researchers are now looking to develop green ML models and architectures by focusing on responsible and sustainable ML systems that will help build a sustainable AI future that goes hand in hand with IoT applications [6].

Green ML for IoT refers to ML research that provides accurate results without incurring higher computational costs, and preferably by reducing them. Green ML focuses on developing and optimizing energy-efficient techniques for performing ML tasks. Although red ML techniques are computationally intensive in many ways, each of them nevertheless may offer an opening for energy-efficient improvements. For example, red ML search can be encouraged to display model accuracy as a function of energy consumption, providing a basis for more data-efficient search in the future. Communicating the computational cost of search, learning, and model execution is a key practice in green ML. Moreover, Green ML techniques can be leveraged in IoT systems to provide energy-efficient solutions to IoT applications. The use of ML in IoT networks with the integration of emerging techniques such as edge computing and softwarization can help provide energy-efficient IoT solutions. In this chapter, we will review several efficiency measures that help optimize energy in ML models for IoT and provide a comprehensive survey that highlights the use of green ML for IoT.

The contributions of this paper are summarized as follows:

- We describe green ML, how to evaluate "greenness" in ML, and why green ML matters.
- We present the background information on the energy management works presented in the literature for green ML and IoT.
- We discuss the energy-efficient infrastructure and inference existing in the literature.

This chapter is organized as follows: In Section 6.2, we present the background information on the Green ML presented in the literature. In Section 6.3, we overview the green IoT. In Section 6.4, we discuss and select some of the energy-efficient infrastructure and inference techniques presented in the literature focusing on the most recent research. In Section 6.5, we present the energy-efficient data usage management technologies for IoT ecosystem. Finally, we conclude this chapter by providing an insight into open research problems and the future of green IoT-enabled ML.

6.2 GREEN MACHINE LEARNING

In this section, we mainly describe green ML, how to evaluate "greenness" in ML, and why green ML matters.

A. **Definition**
Green ML was first introduced by Schwartz [6]; it refers to ML research that yields novel achievements without incurring additional computational costs, and preferably by decreasing resources usages (e.g., computing power, storage space, and energy). Current research and projects related to ML have resulted in a large increase in resource usage. Researchers are seeking to achieve efficiency of their models at the expense of resources usages.

B. **Measures of Efficiency**
To evaluate the efficiency of green ML models, researchers proposed to report the work/resource usage needed to produce ML outputs. These resources include the size of the training dataset and the work needed to train the model, to generate the hyperparameters/parameters during the training experience. When examining the various factors that increase the ML resource usage, we can define three essential factors: the model size (i.e., training dataset), the execution cost (i.e., computational resources during the training and inference), and the hyperparameters/parameters generated (i.e., training rounds and outputs). An experiment's cost can be decomposed into the processing cost of a single example, the size of the dataset, the algorithm complexity, and the number of experiments. Therefore, reducing the cost of each of these tree elements will lead to greener ML.

A comprehensive measure to evaluate the efficiency of green ML models can be expressed as follows:

- **Carbon emission**: Carbon emissions are the first step in measuring environmental impacts [7]. Indeed, the carbon emission can be approximated using the CO_2 equivalents that could be emitted by training or running a model. Nevertheless, this estimation remains inaccurate as the amount of CO_2 emission is very sensitive to the local electrical infrastructure. Additionally, it is complicated for ML contributors to estimate the amount of CO_2 if they do not run their experiments on the same platforms, and there is no well-known platform with a standard metric that estimates this consumption. As a result, it is not suitable for contributors to compare their ML model consumption as they have different infrastructures and run-in different regions.
- **Electricity consumption**: Electricity consumption is related to carbon emissions while being indifferent to time and place. This measure can be related to graphics processing unit (GPUs) that report the amount of electricity each of their cores consumes at any given time, making it easy to assess the total amount of electricity consumed when generating an ML result. However, this measurement is material-dependent and therefore does not allow for a fair comparison between different ML models.
  - **Running time**: Model running time can be adopted to evaluate the efficiency of ML models/algorithms. Assuming all models adhere to the exact same hardware/software parameters, the running time metric is the most appropriate measure for evaluating the training/inference calculations. Since the running time is highly dependent on infrastructure parameters, it cannot be used to compare models running on separate infrastructures.
- **Elapsed real time**: The elapsed time refers to the total amount of time a ML model takes to generate; the shorter the elapsed real time, the more efficient the model is considered. However, this metric is influenced by the used GPU, number of cores, and the other task performed at the same time. Indeed, these factors depend on the computational work performed by the hardware. Although this metric can be very interesting to evaluate a model's efficiency, it is complex to compare the models' performance.
- **The number of parameters**: The number of parameters, i.e., parameters generated and used by the model, is another way to measure model efficiency. Unlike the other measures described above, it is independent of the underlying infrastructure. Additionally, this measure is related to the amount of performed work and the memory used by the model. However, each algorithm makes different use of its parameters; for example, by increasing the depth of the model, the number of parameters gets affected. Therefore, some

models with a similar number of parameters often perform different amounts of work.

- **FLOPs:** Floating-point operations (FLOPs) refer to the amount of work performed by a computational process to run and execute a particular action. This metric is calculated analytically using the basic calculation operations (i.e., multiplication and addition). FLOPs were used as a fair metric to evaluate the energy footprint of models; nevertheless, this metric was rarely used in ML models. Although the FLOPs metrics can be used as an evaluation metric for several ML models, there is a gap between the FLOPs and the running time that need further investigation.

In general, the ML community prefers accurate models with less computation (e.g., reduced number of code lines and low algorithmic complexity). Consequently, it is important to compare equitably the computational resources needed for inference/training of ML models. In this regard, it is recommended to evaluate the FLOPs of ML models as they assess the number of parameters needed to develop a new model.

C. Impact of Green Machine Learning

Green ML can bring ML to better address real-world applications. Over the past years, ML has continued to achieve breakthrough results in various applications, such as speech recognition, computer vision, medical diagnosis, etc. Nevertheless, the cost of energy consumption of ML models is usually overlooked. Nowadays, ML cannot be applied in applications with constrained resources (e.g., smart sensors). In addition, in large-scale applications, resource usage can be multiplied millions of times. As a result, it is essential to further investigate green ML to further ease its integration with resource-constrained applications.

6.3 GREEN IoT AND MACHINE LEARNING

In this section, we first present a brief description of the IoT ecosystem and components. Then, we describe green IoT-enabled machine learning (ML).

A. Internet-of-Things

The present industry is undergoing a paradigm shift from conventional computer- supported industry to smart industry, thanks to recent advances in IoT and other new technologies. During this shift, IoT plays a vital role in bridging the gap between the physical industrial environment and the cyberspace of IT systems.

B. IoT Overview

The IoT is essentially a network of connected smart objects (i.e., things) communicating through wired or wireless communication technologies. It is also defined as the network of physical things that sense, monitor, and control intelligent environments. Areas covered by IoT services include, but are not limited to, supply chain, energy, vehicular, manufacturing, etc. A representative IoT system is composed of the following layered subsystems (from bottom to top), as shown in Figure 6.1.

- **Access Layer**: The access layer contains physical devices that sense and collect data from the physical environment. First, physical devices could either sensor, access points, RFID tags, or wired/ wireless devices that detect the data from the physical environment. Then, other actors, such as actuators and collectors, as well as embedded software running on the devices themselves act on the environment.
- **Communication Layer**: The network communication protocols that allow devices to communicate with each other constitute the center's communication layer. Collected data from the access layer is then transmitted using gateways and base stations. This later form an industrial network to enable communication through the IoT objects. The communication layer is enabled using various communication protocols, such as wired/ wireless –IEEE 802.11, Bluetooth, Infrared, ZigBee, 3G/4G/5G, and LPWAN.
- **Applications Layer**: This layer is the interface between the IoT objects and the networks with which they communicate. It manages the formatting and presentation of data and serves as a bridge between what the IoT device does and the data it transfers to the network. This layer includes a wide number of industrial applications such as industrial IoT (IIoT), Internet of Vehicles (IoV), smart grids, and others.

C. **Green IoT-Enabled Machine Learning**

As IoT becomes more prevalent and the number of IoT devices explodes, the amount of data generated by IoT devices also increases dramatically. ML has emerged as a leading technique to use the IoT data and make IoT application intelligent. Nevertheless, IoT constrained devices (i.e., limited computing power, storage, and energy) have limited the ML adoption. Resource saving, particularly energy saving, has become a priority in IoT-enabled ML applications. The use of green IoT for ML has prompted several researchers to study the impact of green ML for creating green, smart, and sustainable IoT network applications.

The volume of work regarding the use of ML to realize energy conservation/saving is significant. To assist in reducing energy consumption of IoT applications, ML can be employed in energy-saving IoT applications such as smart Energy, smart buildings, smart grid, and others. There are several works that discuss the usage of ML for energy saving in IoT applications [8, 9]. For instance, the authors in

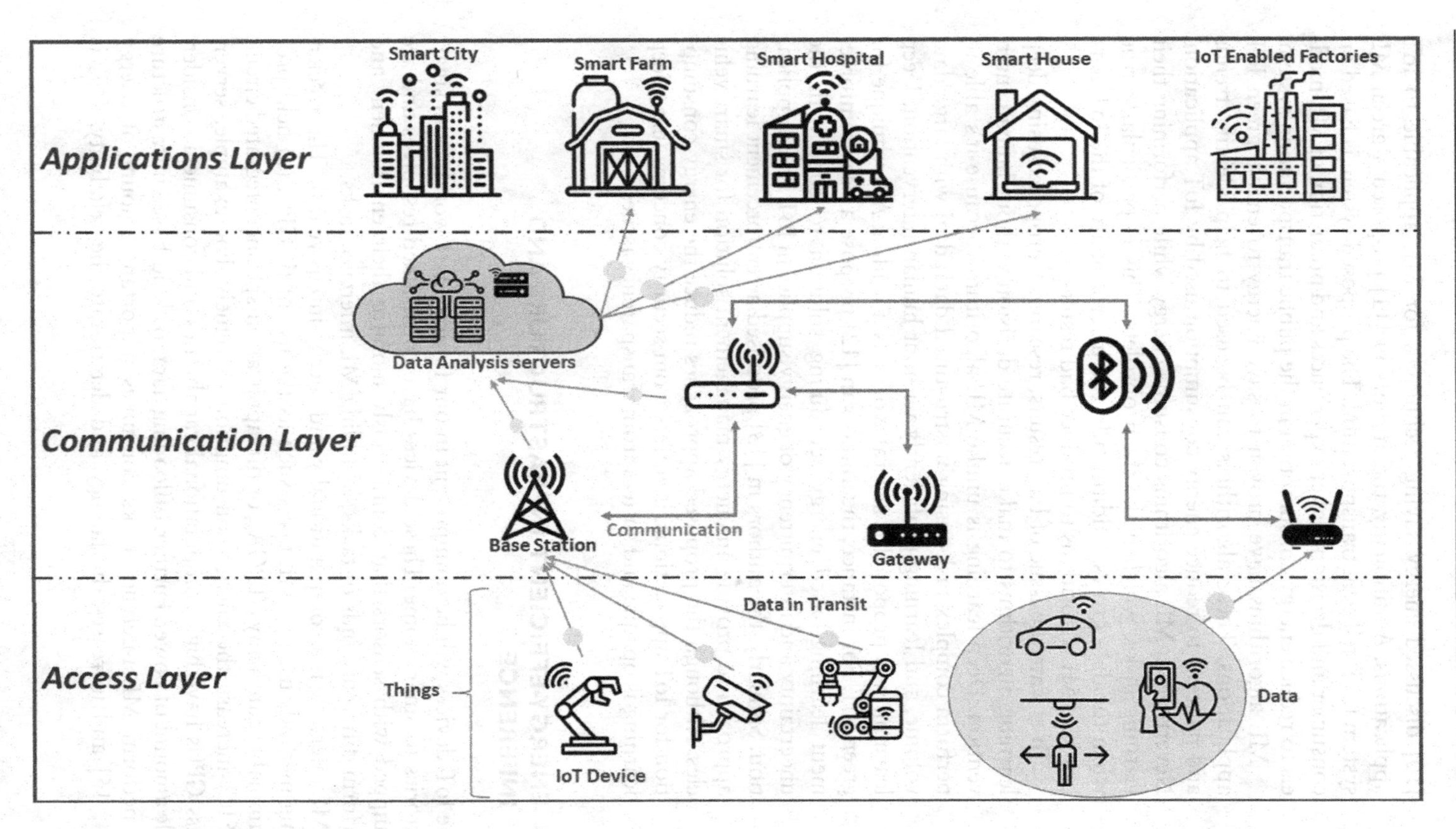

Figure 6.1 IoT ecosystem architecture.

[10] discussed energy saving techniques for ML applicable to IoT applications. Additionally, the authors in [11] proposed a green ML system for intelligent transportation. The proposed system is based on consumer and decision maker experiences and prediction of ML models to reduce energy consumption in the public transportation system.

ML algorithms have proven to save energy to enable green IoT applications. These algorithms can be used to help identify trends and patterns to reduce energy consumption used by IoT application. Nevertheless, ML algorithms consume energy while performing their learning tasks, which make them obsolete. This is particularly true when the data grows. Additionally, the heterogeneity of the IoT data can lead ML algorithms to produce bad results.

To elevate the issue of bad results, researcher proposed using deep learning algorithms to make accurate decision without human intervention. Deep learning is unlike ML algorithms because it's able to perform complex tasks using unstructured data, deal with large data volume, and learning proficiently without human intervention. Deep learning was proposed is several works as a leading AI technique for green IoT. For instance, the authors in [12] propose a deep reinforcement learning-based energy scheduling solution to cope with the uncertainty and intermittency of energy supplies in IoT smart application. Similarly, the authors in [13] proposed a reinforcement learning approach to provide an energy management solution for smart vehicles. Although the proposed approaches reduce the energy consumption for IoT applications, it suffers from security concerns (e.g., deep learning is opaque and suffers from transparency issues).

6.4 ENERGY-EFFICIENT INFRASTRUCTURE AND INFERENCE

The IoT devices will be omnipresent in our future smart world, from smart factories to smart home. These devices have constrained resources and are equipped with sensors that continuously monitor their environment and perform different kinds of tasks, including ML inference tasks.

ML inference is a computationally and energy-intensive task. The accelerating methods used in ML tasks such as the use of (i) GPU, (ii) field programmable gate arrays (FPGA), or (iii) application-specific integrated circuit (ASIC) increase the energy consumption extremely. For example, server-class GPUs have high computational capabilities and consume a considerable amount of power. Further, offloading data to cloud-based infrastructure to perform ML inference tasks consumes important amount of energy [14, 15] and increases the latency and deteriorate the reliability. This is mainly due to the massive data transmission that is generated by the IoT devices and offloaded to the cloud.

A. **Energy-Efficient Green IoT-enabled ML**
To achieve energy-efficient inference, different research efforts have been made to develop energy-efficient methods, including ML-based methods and/or IoT-based methods (algorithmic methods). First, research efforts have been dedicated to developing specialized hardware that will be able to preserve the battery of IoT devices during ML inference tasks. Second, research efforts have been made to optimize the inference algorithms to make them more energy friendly. Third, as our future smart world will use IoT devices as data processing nodes, research efforts have been made to minimize the upload of IoT devices' raw data to the cloud. Therefore, future smart world incorporating green IoT networks must use specialized hardware, optimize the inference algorithms, and use future networking technology including the fog and the edge computing concepts to perform their ML inference tasks [16]. The main idea of fog and edge computing is to process the massive amount of data in the sensors close where it is generated and collected, thereby reducing data transmission energy and latency [17].

To perform their processing tasks, today's IoT devices use microcontrollers, including the advanced reduced instruction set computer machines (ARM) Cortex- M family, which offer milliwatt-order power consumption [17, 18] that is necessary for battery-powered IoT devices to operate properly without excessive battery drain and recharges, which is a necessary condition for future green IoT networks. The downside of microcontrollers is that their computing resources are not sufficient to perform complex processing of many ML inference algorithms [19]. Researchers therefore have proposed different approaches, including parallelism, hardware accelerators, and memory access optimization [20, 21] to achieve the ultralow energy consumption as well as the powerful computation of onboard processing units that will eventually lead to smart and green IoT world.

B. **Inference-Efficient Green IoT-enabled ML**
When optimizing an inference algorithm to perform well on onboard processing units of IoT devices, one must keep in mind of an important trade-off between at least the following three key performance indicators: latency, energy, and accuracy. It was shown that this kind of trade-off exists for many current deep learning and ML inference algorithms. For example, it was shown in [22] that when the deep learning model increases in size, energy consumption increases rapidly for a marginal accuracy increase.

Another alternative to optimize energy efficiency in green IoT-enabled ML networks consists of algorithm-hardware co-optimization. The method in [23] used block-circulant matrices to achieve a fine-grained trade-off between accuracy and compression ratio and it was shown that both the computational and space complexity of the method can be reduced from second-degree polynomial complexity to quasi-linear complexity. Another algorithm-hardware co- optimization technique was proposed in [24] where IoT edge devices perform complex ML inference tasks. Techniques such as deep neural network structure optimization, neuron-synapse optimization and data-type optimization were proposed to optimize the ML algorithms for energy-friendly operations. First, it is important to select the optimal neural network structure for each specific ML task, e.g., select convolutional neural network structure for computer vision tasks, select recurrent neural network structure for sequence processing tasks (or use more specific types of computing such as reservoir computing to reduce the training considerably by only training the output weight of the reservoir). Second, it is also important to select the optimal subset of neurons to activate in order to reduce the number of parameters and thus efficiently reduce the unnecessary operations in the IoT devices' hardware. Finally, reducing the bit-width by optimizing the data type (change the representation of the synapse and neurons values from single precision floating point to dynamic fixed points for example) is of a great importance.

The acceleration of ML inference tasks in green IoT networks is of great importance to reduce energy consumption. To accelerate ML inference in green IoT networks, low-rank approximation techniques have been proposed and have been shown to perform well due to their mathematical theory and efficient implementations. Low-rank decomposition works by decomposing the neural network into a network that is compact and has low-rank layers [25]. Further, pruning techniques, such as weight pruning and filter pruning techniques, have shown to be effective in reducing the energy consumption and preserving accuracy. Both low-rank decomposition and pruning can be combined to obtain higher model compression and reduce energy consumption during inferencing [26]. Knowledge distillation is also a technique that can train small neural networks that capable to mimic large ones [27]. A student-teacher approach can be used to train a smaller neural network, given state-of-the-art larger network. The small network, after networks' parameters quantization, will fit on battery powered IoT devices and thus will be compatible with green IoT networks. Knowledge distillation can also be combined with model early exit to accelerate inference through augmenting the deep network architecture with additional branch classifiers [28, 29].

6.5 EFFICIENT DATA USAGE IN GREEN IoT-ENABLED ML

In this section, we first overview methods for efficient data usage in green IoT-enabled Ml. Then, we present some limitations of these methods and some future work directions.

A. **Energy-Efficient Green IoT-enabled ML**
The IoT comprises a large number of heterogeneous and interconnected objects (e.g., devices) that exchange data over the network [30, 31]. To enable energy-efficient data usage in green IoT-enabled ML, we briefly review some methods for reducing inference costs in ML models [32]. These methods include model pruning, quantization, and knowledge distillation.

Model pruning is one model compression technique that optimizes in real time the model inference for resource-constrained devices [33]. This technique consists of removing the nonsignificant weights from the sampling. Consequently, it improves the training/inference speed, model generalization, and data samples. Model pruning has motivated several researchers to remove insignificant parameters from the original network to generate a smaller network that can preserve the accuracy of the original model. Nevertheless, model pruning still suffers from some limitations that can affect model performance. Model pruning needs to iteratively record weights/parameters and retrain the model over many iterations which are computationally intensive. Therefore, this method should be further investigated to enable efficient pruning algorithms.

Model quantization consists of compressing the original model size by reducing the number of bits [34]. Model quantization was used in several ML models such as CNN where a 32-bit encoded CNNs can be transformed into 8-bit encoded CNNs, thus considerably reducing storage and computing costs. Model quantization includes two types of methods: deterministic quantization, where the deterministic correspondence is realized between the quantified value and the real value and stochastic quantization, where the quantified value is extracted from a discrete random distribution. In general, post-training quantization is most straightforward; it consists of applying quantization on a trained model to reduce inference costs. Nevertheless, this method can affect the model performance. To address this issue, quantization-aware training can be leveraged by refining the quantified model before inference.

Knowledge distillation consists of using the knowledge of a well-trained model, refer to as a "teacher," to assess training a new model, refer to as a "student" [35]. The knowledge distillation method reduces the gap between the knowledge in the teacher's model and the knowledge in the student's model. This method can be efficient in

IoT-enabled ML applications to have a competitive model but smaller and suitable with resource constrained applications. Although this method is easy to implement, it is sensitive to the discrepancy between the teacher's model and the student's model.

B. Limitations and Future Work

Through our work on this chapter, we have identified several future directions and open issues that need further investigation:

- The green IoT ecosystem (i.e., hardware and software) should be redesigned to reduce its energy consumption. For instance, hardware based can be energy saving sensors, servers, and processor, while software based can be cloud based, virtualization, and others.
- Given the scarcity of the communication resources in the IoT context, enabling green IoT enable ML is complex. In fact, the generated traffic of IoT objects grows linearly with the number of ML model size. Therefore, it is essential to use IoT communication resources wisely and to think about implementing new perceptive learning methods to enable green ML for IoT.
- The usage of multiple machine learning model or ensemble learning could reduce the resource usage for green IoT. Ensemble learning involves combining the predictions from multiple contributing models. Ensemble learning could help reduce the variance components and therefore reduce the resource usage.

6.6 CONCLUSION

Green ML in the IoT will be an important research direction to help overcome the inclusiveness challenges of current IoT-enabled ML applications. It is expected that green ML will reduce the computational expense while preserving the minimal model performance to move IoT application in a cognitive plausible direction. With recent advances in ML, the research community has developed several efficient energy, inference, and data usage suitable with the IoT constrained systems. Considering the goal is to deploy ML models on real-world devices with constrained resources, green ML needs further investigation to achieve this goal.

We have highlighted several potential research directions for green ML in the IoT context, but still, some open questions deserve to be explored in depth. For instance, the model parameters that need to be optimized to leverage green ML models cannot be standardized between all the models. Indeed, the number of parameters for training and inference cannot be fixed between all ML models. Additionally, it is complex to develop green ML models with high performance to enable efficient zero-shot learning suitable for IoT sensitive applications (i.e., smart hospitals) like a human. Furthermore, data usage is still an open research subject, particularly in the IoT context. ML models constantly learn using newly received data, how models store knowledge, and how to make models achieve lifelong learning without forgetting the knowledge acquired during training is still a complex problem to solve.

REFERENCES

[1] F. Firouzi, B. Farahani, F. Ye, and M. Barzegari, 2020. Machine Learning for IoT. In *Intelligent Internet of Things: From Device to Fog and Cloud*, F. Firouzi, K. Chakrabarty, and S. Nassif, Eds. Cham: Springer International Publishing, 243–313.

[2] H. Moudoud, L. Khoukhi, and S. Cherkaoui, 2020. "Prediction and Detection of FDIA and DDoS Attacks in 5G Enabled IoT," in *IEEE Network*, vol. 35, no. 2, 194–201.

[3] H. Moudoud, Z. Mlika, L. Khoukhi, and S. Cherkaoui, 2022. "Detection and Prediction of FDI Attacks in IoT Systems via Hidden Markov Model," in *IEEE Transactions on Network Science and Engineering*, vol. 9, no. 5, 2978–2990.

[4] D. Patterson, J. Gonzalez, Q. Le, C. Liang, L. Munguia, D. Rothchild, D. So, M. Texier, and J. Dean, 2021. Carbon Emissions and Large Neural Network Training, arXiv:2104.10350.

[5] R. Schwartz, J. Dodge, N. A. Smith, and O. Etzioni, 2021. Green AI, arXiv:1907.10597.

[6] A. Lacoste, A. Luccioni, V. Schmidt, and T. Dandres, 2019. Quantifying the Carbon Emissions of Machine Learning. CoRR, abs/1910.09700.

[7] Y. Zhou, F. R. Yu, J. Chen, and Y. Kuo, 2019. Robust Energy-Efficient Resource Allocation for IoT-Powered Cyber-Physical-Social Smart Systems with Virtualization, in *IEEE Internet of Things Journal*, vol. 6, no. 2, 2413–2426.

[8] L. Brian, T. J. Cook, and D.J. Cook, 2016. Activity-Aware Energy-Efficient Automation of Smart Buildings, in Energies, vol. 9, no. 8, 624.

[9] L. Paris, and M. H. Anisi, 2019. An Energy-Efficient Predictive Model for Object Tracking Sensor Networks, *2019 IEEE 5th World Forum on Internet of Things (WF-IoT)*, pp. 263–268.

[10] Y. Chen, A. Ardila-Gomez, and G. Frame, 2017. Achieving Energy Savings by Intelligent Transportation Systems Investments in the Context of Smart Cities, in *Transportation Research Part D: Transport and Environment*, vol. 54, 381–396.

[11] Y. Liu, C. Yang, L. Jiang, S. Xie, and Y. Zhang, 2019. Intelligent Edge Computing for IoT-Based Energy Management in Smart Cities, in *IEEE Network*, vol. 33, no. 2, 111–117.

[12] Y. Hu, W. Li, K. Xu, T. Zahid, F. Qin, and Li, C., 2018. Energy Management Strategy for a Hybrid Electric Vehicle Based on Deep Reinforcement Learning, in *Applied Sciences*, vol. 8, no. 2, 187.

[13] M. Mohammad, D. Fong, and S. Ghiasi, 2016. Fast and Energy-Efficient CNN Inference on IoT Devices. arXiv preprint arXiv:1611.07151.

[14] F. Samie, L. Bauer, and J. Henkel, 2019. From Cloud Down to Things: An Overview of Machine Learning in Internet of Things, in *IEEE Internet Things J.*, vol. 6, no. 3, 4921–4934.

[15] X. Wang, M. Magno, L. Cavigelli, and L. Benini, 2020. FANN-on- MCU: An Open-Source Toolkit for Energy-Efficient Neural Network Inference at the Edge of the Internet of Things, in *IEEE Internet of Things Journal*, vol. 7, no. 5, 4403–4417.
[16] A. Rahimi, P. Kanerva, and J. M. Rabaey, 2016. A Robust and Energy-Efficient Classifier Using Brain-Inspired Hyperdimensional Computing, in *Proc. ACM/IEEE Int. Symp. Low Power Electron. Design*, pp. 64–69.
[17] M. Magno, M. Pritz, P. Mayer, and L. Benini, 2017. DeepEmote: Towards Multi-Layer Neural Networks in a Low Power Wearable Multi-Sensors Bracelet, in *Proc. IEEE Int. Workshop Adv. Sensors Interfaces*, pp. 32–37.
[18] M. Qin, L. Chen, N. Zhao, Y. Chen, F. R. Yu, and G. Wei, 2019. Powerconstrained Edge Computing with Maximum Processing Capacity for IoT Networks, in *IEEE Internet Things J.*, vol. 6, no. 3, 4330–4343.
[19] L. Cavigelli, and L. Benini, 2017. Origami: A 803-GOp/s/W Convolutional Network Accelerator, *IEEE Trans. Circuits Syst. Video Technol.*, vol. 27, no. 11, 2461–2475.
[20] R. Andri, L. Cavigelli, D. Rossi, and L. Benini, 2019. HyperDrive: A Multichip Systolically Scalable Binary-Weight CNN Inference Engine, *IEEE J. Emerg. Sel. Topics Circuits Syst.*, vol. 9, no. 2, 309–322.
[21] L. Cavigelli, G. Rutishauser, and L. Benini, 2019. EBPC: Extended Bitplane Compression for Deep Neural Network Inference and Training Accelerators, *IEEE J. Emerg. Sel. Topics Circuits Syst.*, vol. 9, no. 4, 723–734.
[22] Walid A. Hanafy, Tergel Molom-Ochir, and Rohan Shenoy, 2021. Design Considerations for Energy-efficient Inference on Edge Devices, in *Proceedings of the Twelfth ACM International Conference on Future Energy Systems (e-Energy '21). Association for Computing Machinery*, New York, NY, USA, pp. 302–308.
[23] Yanzhi Wang et al., 2018. Towards Ultra-High Performance and Energy Efficiency of Deep Learning Systems: An Algorithm-Hardware Co-Optimization Framework, in *Proceedings of the AAAI Conference on Artificial Intelligence*, vol. 32, no. 1.
[24] J. Lee, S. Kang, J. Lee, D. Shin, D. Han, and H. -J. Yoo, 2020. The Hardware and Algorithm Co-Design for Energy-Efficient DNN Processor on Edge/Mobile Devices, in *IEEE Transactions on Circuits and Systems I: Regular Papers*, vol. 67, no. 10, 3458–3470.
[25] Y. Xu et al., 2019. Trained Rank Pruning for Efficient Deep Neural Networks, in *2019 Fifth Workshop on Energy Efficient Machine Learning and Cognitive Computing - NeurIPS Edition (EMC2-NIPS)*, pp. 14–17.
[26] S. Goyal, A. Roy Choudhury, and V. Sharma, 2019. Compression of Deep Neural Networks by Combining Pruning and Low Rank Decomposition, in *2019 IEEE International Parallel and Distributed Processing Symposium Workshops (IPDPSW)*, 952–958.
[27] G. Cerutti, R. Prasad, A. Brutti and E. Farella, 2020. Compact Recurrent Neural Networks for Acoustic Event Detection on Low-Energy Low-Complexity Platforms, in *IEEE Journal of Selected Topics in Signal Processing*, vol. 14, no. 4, 654–664.
[28] W. Fang, F. Xue, Y. Ding, N. Xiong, and V. C. M. Leung, 2021. EdgeKE: An On-Demand Deep Learning IoT System for Cognitive Big Data on Industrial Edge Devices, in *IEEE Transactions on Industrial Informatics*, vol. 17, no. 9, 6144–6152.

[29] T. Surat, B. McDanel, and H. Kung, 2016. Branchynet: Fast Inference via Early Exiting from Deep Neural Networks, in *2016 23rd International Conference on Pattern Recognition (ICPR)*. IEEE.
[30] H. Moudoud, S. Cherkaoui, and L. Khoukhi, 2021. Towards a Scalable and Trustworthy Blockchain: IoT Use Case, in *ICC 2021 - IEEE International Conference on Communications*, 1–6.
[31] H. Moudoud, S. Cherkaoui, and L. Khoukhi, 2019. An IoT Blockchain Architecture Using Oracles and Smart Contracts: the Use-Case of a Food Supply Chain, in *2019 IEEE 30th Annual International Symposium on Personal, Indoor and Mobile Radio Communications (PIMRC)*, 1–6.
[32] A. Lacoste, A. Luccioni, V. Schmidt, and T. Dandres, 2019. Quantifying the Carbon Emissions of Machine Learning, in *Proceedings of the Climate Change AI Workshop*.
[33] S. Hanson, and L. Y. Pratt, 1988. Comparing Biases for Minimal Network Construction with Back-Propagation In D. S. Touretzky (Ed.), *Advances in neural information processing systems (NIPS)*, vol. 1 (pp. 177–185). San Mateo, CA: Morgan Kaufmann.
[34] V. Vanhoucke, A. Senior, and M.Z. Mao. 2011. Improving the Speed of Neural Networks on CPUs. In *Proc. Deep Learning and Unsupervised Feature Learning NIPS Workshop*.
[35] G. Hinton, O. Vinyals, and J. Dean. 2015. Distilling the Knowledge in a Neural Network. arXiv preprint arXiv:1503.02531, 2(7).

Chapter 7

Green machine learning protocols for machine-to-machine communication

Sreenivasa Reddy Yeduri, Sindhusha Jeeru and Linga Reddy Cenkeramaddi
University of Agder, Grimstad, Norway

7.1 INTRODUCTION

In this section, we introduce the cellular internet-of-things and the latest advancements in it. Thereafter, we introduce massive machine-to-machine communication. Finally, an introduction to green machine learning models is presented.

7.1.1 Cellular Internet-of-Things

The combination of evolving wireless technologies and the Internet-of-Things (IoT) applied areas like industrial automation, connected cars, and smart grid technologies has created a platform for the development of smart and connected societies in the near future, as shown in Figure 7.1. This platform envisions enhancing the overall lifestyle and quality of people's lives by integrating numerous smart devices (predicted to be over 125 billion by the year 2030 as forecasted by Information Handling Services (IHS) Markit) into intelligent industry processes, people, and society [1]. Recently, there have been several developments in areas like licensed and unlicensed cellular technologies that support connected IoT devices. In particular, the licensed technologies include Long-Term Evolution for machine-to-machine Communications (LTE-M) and Narrow-Band IoT (NB-IoT). In contrast, the unlicensed technologies include ZigBee, Wi-Fi, and LoRa [2]. Among all, cellular technologies are considered to be one of the best because they provide a wide range of coverage area, good Quality of Service (QoS) provisioning in terms of performance, and tight coordination. Hereby, cellular IoT has been one of the major focuses in the research sector.

7.1.1.1 Latest advancements in cellular IoT

Recently, cellular IoT has attracted drastic attention from various research collaborations like academia, industrial research, and standardization bodies to integrate IoT devices into already existing cellular infrastructure. The upcoming cellular networks like 5G and beyond seek to emerge with diverse

DOI: 10.1201/9781003230427-7

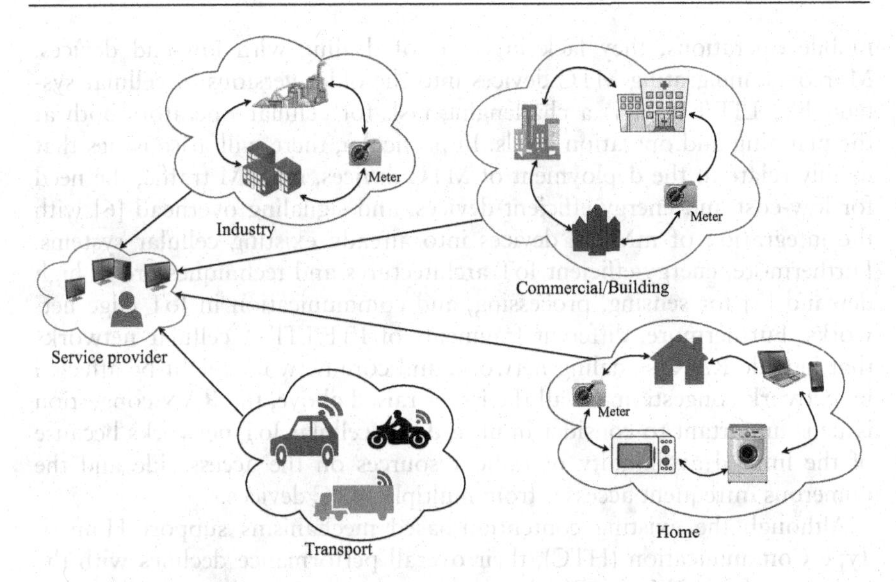

Figure 7.1 Applications of cellular Internet-of-Things.

telecommunication services in the future, and these services are categorized into three different classes by the international telecommunication union radiocommunication (ITUR) sector [3]. The classes are namely (i) massive Machine Type Communications (mM2M), (ii) enhanced Mobile Broadband (eMBB), and (iii) Ultra-Reliable and Low Latency Communications (URLLC).

Among these classes, the eMBB comes up with high data rate requirements for 5G systems, whereas the mM2M provides large scalability for connecting multiple devices, of the order of one million devices per square kilometer with diverse QoS requirements [4]. Additionally, URLLC seeks to ensure a robust connection with very low delay. The primary task while dealing with an mM2M case is to handle numerous devices with an inadequate number of radio resources. On the other hand, URLLC has to deal with providing an extremely reliable network in the order of 99.999% within a very short span of 1 ms [5]. Among the abovementioned scenarios, mM2M has to handle a few non-conventional challenges that include managing highly dynamic and infrequent (sporadic) traffic, congestion-related issues in Radio Access Network (RAN), QoS provisioning, and large signaling overhead. Therefore, the main focus of this chapter is on mM2M.

7.1.1.2 Challenges in cellular IoT

Although there are various advantages to taking centralized cellular systems into consideration, which mainly include allowance of broad coverage area, tight time synchronization for wireless networks (Tiny-Sync), and on-hand mobile operations, they lack in terms of dealing with low-end devices. Moreover, integrating MTC devices into the older versions of cellular systems like LTE/LTE-A is a challenging task for cellular operators both at the planning and operation levels. In particular, there will arise issues that mainly relate to the deployment of MTC devices, mM2M traffic, the need for low-cost and energy-efficient devices, and signaling overhead [6] with the integration of mM2M devices into already existing cellular systems. Furthermore, energy-efficient IoT architectures and techniques are in high demand [7] for sensing, processing, and communication in IoT edge networks. Furthermore, different fragments of LTE/LTE-A cellular networks that include RAN, signaling network, and core network might be affected by network congestion. Of all the issues raised above, the RAN congestion issue is important to consider in ultra-dense cellular IoT networks because of the limited availability of radio resources on the access side and the numerous infrequent accesses from multiple MTC devices.

Although the existing contention-based mechanisms support Human-Type Communication (HTC), their overall performance declines with the inclusion of mM2M scenario due to the numerous sporadic access requests. There will also arise an issue related to a limited number of preambles in LTE-based systems as it results in the selection of the same preamble at the same time, which in turn results in numerous collisions in the access network. The access attempts in the mM2M scenario are supposed to be dynamic and sporadic, which leads to high traffic during data transmission in both traffic and access channels. This is beyond the capacity to be handled by the IoT access network, which in turn leads to unavoidable congestion in the IoT access network [8]. IoT devices come with very high signaling overhead [6, 9] per data packet, which is a critical issue that has to be considered. Taking the numerous challenges into consideration, it is very important to use emerging tools such as machine learning (ML) to work on efficient signaling reduction and suitable transmission scheduling techniques in ultra-dense schemas.

7.1.2 RACH mechanism

Cellular IoT is a type of technology where millions of MTC devices connect to a base station (BS) for transmitting data without any human intervention. Initially, all MTC devices stay in a radio resource control (RRC)-idle state where neither transmission of data nor connection to BS takes place. Whenever an MTC device needs to exchange information, it seeks to connect to the BS after waking up. The mechanism in which an MTC device makes a connection moving from an RRC-idle state to an RRC-connect state is titled the RACH mechanism.

The RACH mechanism is performed with a four-message exchange as described in Figure 7.2, wherein four unique message exchanges take place between each MTC device and the BS. At an initial stage, BS broadcasts a

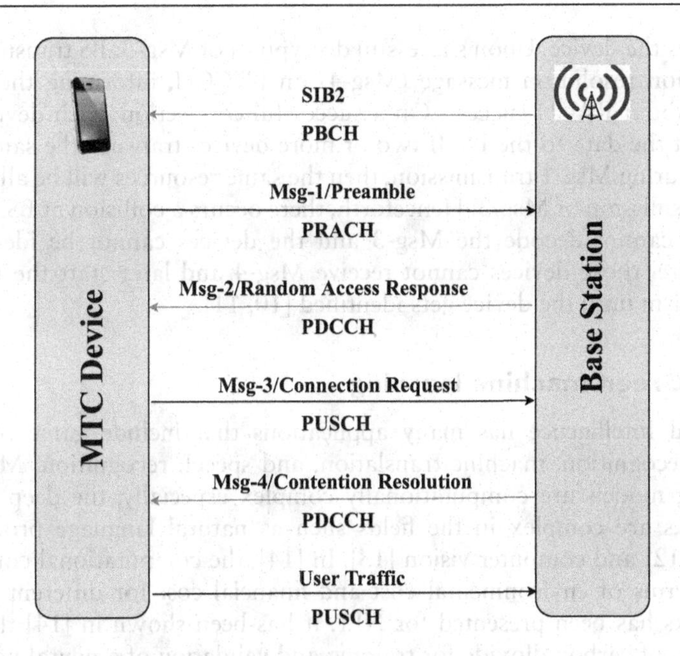

Figure 7.2 The random access mechanism.

system information block-2 (SIB-2) message over a channel called the Physical Broadcast Channel (PBCH) to all the MTC devices to initiate the RACH mechanism. MTC devices that are waiting to connect, successfully decode the SIB-2 message and respond with a randomly picked preamble from 54 orthogonal preambles, and transmit as Message-1 (Msg-1). The Msg-1 is transmitted over a channel called the shared Physical RACH (PRACH).

SIB-2 is composed of group-A and group-B blocks, which define the length of the preambles that have to be transmitted from the MTC device. Henceforth, the length of the preamble that is transmitted as Msg-1 depends on the lengths defined in the SIB-2 message. There is a probability that different MTC devices can transmit the same or different preamble to the BS. Except for the preamble, Msg-1 doesn't contain the identity of the device. Therefore, BS cannot identify the identity of the device from which the particular preamble has been generated and transmitted. As soon as the BS receives the Msg-1, it decodes the preamble and responds with Msg-2, called a random access response (RAR), over a channel called the Physical Downlink Control Channel (PDCCH). RAR contains information about the resource blocks allocated for the transmission of Msg-3. On successful receipt of Msg-2, MTC device responds with a connection request message (Msg-3) transmitted over a Physical Uplink Shared Channel (PUSCH) which carries the device identity. If a unique preamble is activated by an MTC device during Msg-1 transmission, then BS successfully decodes and identifies the device. Upon successful decryption of Msg-3, BS transmits the contention resolution message (Msg-4) on PDCCH, informing the MTC device about RACH success. On a successful connection, each device can transmit the data to the BS. If two or more devices transmit the same preamble during Msg-1 transmission, then the same resources will be allocated for transmission of Msg-3. Henceforth, there occurs a collision at BS, which in turn cannot decode the Msg-3 and the devices cannot be identified. Therefore, those devices cannot receive Msg-4 and later start the RACH mechanism until the device gets identified [10, 11].

7.1.3 Green machine learning

Artificial Intelligence has many applications that include game playing, object recognition, machine translation, and speech recognition. Machine learning models are computationally complex especially, the deep neural networks are complex in the fields such as natural language processing (NLP) [12] and computer vision [13]. In [14], the computational complexity in terms of environmental cost and financial cost for different neural networks has been presented for NLP. It has been shown in [14] that the emission of carbon dioxide for training and validation of a neural network is five times more than the lifetime emission of an average American car. Thus, green ML models are the emerging research models which are the main contribution of this chapter.

7.1.3.1 Definition

Green artificial intelligence (AI), proposed in [15], is a learning model that is expected to result in comparable results with less computational complexity for training and validation. While the Red AI results in higher training and validation accuracies with higher computational cost, the green AI results in reasonable results with fewer computations.

7.1.3.2 Measures

In green AI, the computational metrics are very important. Nowadays, there is a lack of proper measures for computations as it depends on many factors such as training examples, model sizes, and so on. Thus, we need a comprehensive measure. Following are some of the measures.

- **Measures of efficiency**: It is a measure of the amount of work required for generating a result. Specifically, it is defined as the amount of work required for training the dataset or tuning the hyperparameters. In the end, the cost of an experiment is measured as the work done in processing an example, which further depends on the size of the dataset and the number of experiments. Reducing the work done in each of these steps results in an AI which is more green. The work done can be measured using the amount of carbon emission, electricity usage, elapsed real-time, and the number of parameters:

- **Carbon emission:** The amount of carbon released for training or executing a model. The amount of carbon released depends on the local electricity infrastructure. Since the consumed electricity depends on the location and time, it is not recommended for comparison.
- **Electricity usage:** The amount of electricity consumed is directly correlated to the amount of carbon released. Further, the GPU consumed energy depends on the energy consumed by the cores. Further, the amount of electricity consumed depends on the hardware which depends on the location.
- **Elapsed real time:** It is the total time for generating the result of AI which is a suitable measure for efficiency. With other parameters being fixed, the faster model needs less work. However, the elapsed time depends on the other models running on the machine and the number of cores used.
- **Number of parameters:** Unlike the other parameters described above, the number of parameters does not depend on the hardware used. It directly correlated with the memory to be used by the model. Different models use the number of parameters differently, for instance, it is used for making the model deeper or wider. Thus, different models with same number of parameters require different amount of work.

7.1.4 The necessity and associated challenges for ML in mM2M networks

Moving from one generation of the cellular network to the upcoming one, the number of parameters that have to be configured increases exponentially. This leads to challenging tasks like optimization of cellular networks in both wireless and dynamic environments [16, 17]. Existing wireless systems use link adaptation techniques that are familiar with different physical layer parameters, which include coding schemes based on the reliable communication link. This would not be advantageous in ultra-dense cellular IoT networks. Also, the adaption techniques rely on the prediction of wireless link reliability through metrics like Packet Error Rate (PER). This prediction becomes complicated as it involves the use of wideband signals, multiple antennas, and various advanced signal processing algorithms [18]. Moreover, using conventional signal processing tools makes it even more difficult to predict the accurate PER. In addition to that, with the involvement of numerous environmental parameters like signal power, information of channel state, non-Gaussian noise effect, noise variance, and transceiver hardware impairments, the achievement of an efficient adaption link becomes even more complicated [19]. The severity of the problem is going to increase with the ultra-dense networks that involve numerous devices and system parameters.

From the above discussion, it is clear that there is a need for the development of context-aware adaptive communication protocols that include battery levels of the devices, quality of adaption link, environmental parameters status, geospatial information, network states, and energy consumption, to take optimized decisions in a wireless environment [20]. The idea of self-handling is a challenging task when it comes to configuration and optimization in ultra-dense cellular networks. This is because one needs to observe dynamic variations in the environment, learn the uncertainties, and plan to effectively configure the involved network parameters, which is a challenging task. Therefore the emerging ML-based techniques seem promising as they can be functioned to learn the system uncertainties based on the parameters, and can be programmed to predict the results for future investigation [17]. The future ultra-dense cellular network is supposed to deal with all the included distributed systems to increase the overall system performance. This can be achieved with ML-based techniques used with the collaborative edge-cloud processing platform [21], as the use of global network knowledge is enabled at the edge-side and also the functionality of coordination is enabled among different distributed systems. Recently, ML-based techniques have gained wide attention to handle the issues occurring in dynamic wireless environments [21, 22]. There were also some existing works regarding MTC environments that were published for the study of applying various ML-based techniques to learn various cellular network parameters [22–25].

The remainder of this chapter is organized as follows. The motivation for the green machine learning for RACH mechanism in mMTC is described in Section 7.2. A survey on the existing learning-based techniques and their measures are elaborated in Section 7.3. Section 7.4 presents the experimental results to compare the performance of the ML and non-ML-based approaches. Section 7.6 concludes this chapter with possible future directions in Section 7.7.

7.2 MOTIVATION

In this chapter, we start with the contention resolution in MTC followed by the motivation for the need for machine learning models.

Contention in the RACH is defined as the number of devices that contend for RACH success in a time slot. With N being the number of accesses and K being the available preambles, the number of access successes is obtained as

$$S = N(1-1/K)^{N}$$

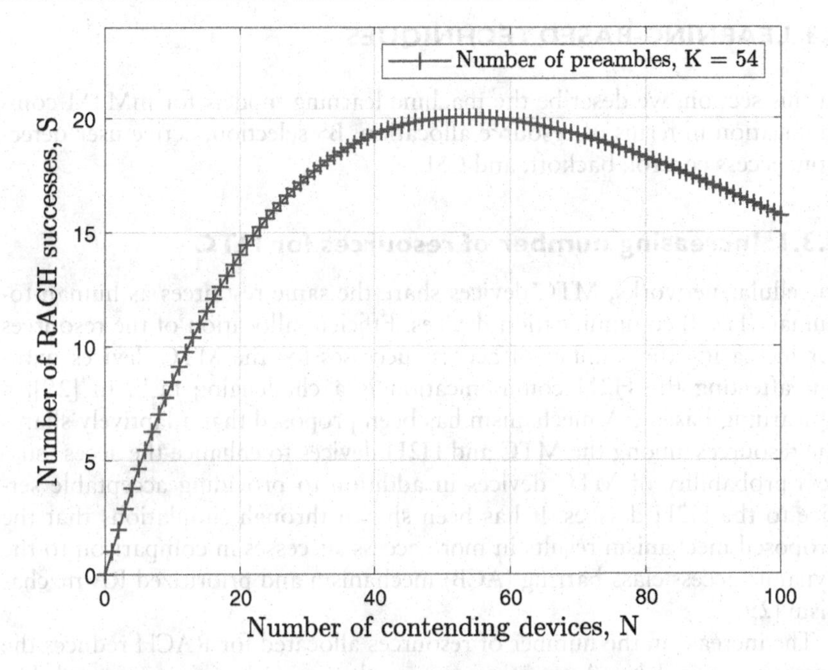

Figure 7.3 Access successes vs number of contending devices.

Figure 7.3 shows access successes vs number of contending devices plot for $K = 54$. It is noticed from Figure 7.3 that the access successes increase exponentially with an increasing number of accesses and reach a maximum at $K = N$. Further increase in the number of accesses reduces the access successes. Thus, in addition to increasing the allocated resources, one should control the number of accesses in a radio frame to avoid channel congestion and improve access successes.

The exponential growth of mM2M devices increases the congestion in the radio access network. To mitigate this effect, 3GPP has proposed an EAB mechanism that controls the number of accesses in a radio frame so as to increase the number of RACH successes [26]. There are several theoretical mechanisms in the literature to enhance 3GPP-EAB's performance. Recently, a successive interference cancellation (SIC)-based RACH mechanism has been proposed in [10, 27] that utilizes the inter-slot SIC at the BS to improve the number of RACH successes. Thereafter, a SIC-NORA mechanism has been proposed in [11] that utilizes both intra-slot and inter-slot SIC at the BS in order to further enhance the success. However, these mechanisms need to solve highly complex and long-term optimization problems to configure the network parameters for future transmissions based on the traffic in the current transmissions. Machine learning (ML)-based approaches are more efficient in solving these high-complexity optimization problems in comparison to non-ML-based approaches [28].

7.3 LEARNING-BASED TECHNIQUES

In this section, we describe the machine learning models for mM2M communication in terms of resource allocation, BS selection, active user detection, access control, backoff, and CSI.

7.3.1 Increasing number of resources for MTC

In cellular networks, MTC devices share the same resources as human-to-human (H2H) communication devices. Efficient allocation of the resources for increasing the number of access successes for the MTC devices without affecting the H2H communication is a challenging task. In [29], a Q-learning-based RA mechanism has been proposed that adaptively shares the resources among the MTC and H2H devices to enhance the access success probability of MTC devices in addition to providing acceptable service to the H2H devices. It has been shown through simulations that the proposed mechanism results in more access successes in comparison to the dynamic access class barring (ACB) mechanism and prioritized RA mechanism [29].

The increase in the number of resources allocated for RACH reduces the average access delay. A situation-aware adaptive Q-learning approach has been considered in [30] to adaptively control the number of resources allocated for both data transmission and RACH. This, in turn, adds the unused data transmission resources to the RACH resources, increasing the total allocated resources and lowering the average access latency. It has been shown through results that a learning rate and a discount factor of 1 and 0.8, respectively, are the best parameters for the proposed approach. Further, the reward has been chosen to be the difference between the present waiting time and the waiting time before [30]. The collaborative Q-learning algorithm proposed in [30] is model-free and does not have planning. Thus, a model-based approach needs to be considered to make it a green model.

7.3.2 BS selection

In [31], a learning-based approach has been proposed for the selection of the appropriate BS in order to avoid congestion in the network and reduce packet delivery delay. Further, in [32], the authors have proposed a Q-learning-based unsupervised learning algorithm for the MTC devices to select the appropriate BS based on the QoS parameters in the presence of dynamic traffic in the network. It has been shown through simulations that the proposed algorithm improves the QoS and helps in the self-organization of the RA in a cell. A reinforcement learning algorithm has been proposed in [33] that works with value-difference based exploration for solving the problem of BS selection. Here, a Markov decision process is used to formulate the problem. It has been shown through the simulations that the proposed algorithm balances the traffic load and results in better performance in terms of blocking probability with an acceptable transmission rate.

7.3.3 Active user detection (AUD)

An end-to-end deep neural network (DNN) model has been employed in [34] to design the spreading sequence and to detect the active users for the massive number of MTC devices accessing the BS using non-orthogonal spreading sequences. In the employed DNN model, the transmitter side is modeled using the spreading network, and the active devices are estimated using an active user detection network. Through numerical results, it has been shown that the spreading sequence obtained from the proposed approach results in greater performance in terms of active user detection in comparison to the compressive sensing-based approach. A fast DL-based method has been proposed in [35] for compressive sensing-based multiuser detection in mM2M systems. A neural network algorithm named the deep learning (DL)-modified orthogonal approximate message passing-user identification (DL-mOAMP-UI) algorithm has been presented in [36] to minimize the detection probability of the missed user under a pre-determined false alarm probability. The computational complexity of the proposed method in [36] is high which needs to be reduced to make it green.

The grant-free (GF) access and NOMA systems are commonly used to support a massive number of transmissions. Active device identification is required at the BS as the devices transmit the data without scheduling in GF-NOMA systems. In addition, channel estimation is also a challenging task as multiple channels are superimposed in the same resource block. In [37], DL-based AUD and CE have been proposed for GF-NOMA systems. Here, long-short-term memory (LSTM) has been incorporated for mapping the indices of active devices, the received NOMA signals, and associated channels. It has been shown through numerical results that the proposed method is constructive with AUD and CE for mM2M [37]. The run time for training of the proposed method is 7.41×10^4 which is extremely high. This needs to be reduced to make it a green model.

7.3.4 Access control at the BS

An access mechanism based on Q-learning has been proposed in [38] that coordinates among the MTC devices to avoid collisions among them without the involvement of a central entity. After the convergence of learning, devices get a unique RA slot for transmission. It has been shown through simulations that the proposed approach improves the RACH throughput by 55% and also overcomes the negative impact of M2M traffic on the H2H traffic [38]. In [39], a hierarchical learning algorithm has been proposed at the BS that enables the decision control at the MTC device based on the received common control message (CCM). A Q-learning-based access control mechanism has been proposed in [40] to solve the overload problem when a huge number of MTC devices try to access the system using contention-based RACH. A survey of the learning-based ACB methods has been presented in [41] by classifying them into adaptive barring factor (ABF) ACB solutions and ABF and barring time ACB solutions. It has been shown in [41] that the ACB solutions based on learning adaptively tune the access request traffic (ART) and ABF improve the access delay (AD) along with maintaining the baseline-ACB's best access success probability (ASP) performance. However, the above method consumes more energy. Thus, an approach that adaptively tunes both ABF and ART results in lower energy consumption with similar AD and ASP performances [41]. A probabilistic method with a combination of multiple context metrics through machine learning has been proposed in [42] for vehicular MTC that is shown to achieve a higher average data rate of 164% in comparison to regular periodic transmission. The proposed work in [42] focused on the improvement of the accuracy which needs a lot of training and consumes high resources for training. To make it green, the model has to be designed to train with lower data and should result in higher accuracy. A deep reinforcement learning algorithm has been presented in [43] to dynamically control the ACB factor in a homogeneous priority network. Thereafter, the algorithm has been extended for heterogeneous MTC devices with different quality of service requirements. Through simulations, it has been shown that the proposed algorithm's performance coincides with the theoretical optimum for homogeneous networks, and for heterogeneous networks, it results in higher access probability and lower delay [43]. A non-orthogonal-multiple-access (NOMA) and Q-learning-based distributed method has been proposed in [44] to dynamically allocate the RA slots to the MTC devices. It has been shown through numerical results that the proposed method in [44] results in higher network throughput compared to the other methods.

In [45], an online hierarchical stochastic learning algorithm has been proposed to make the access control decision at the MTC devices separately based on the Q-factors. This has been done with the assistance of CCM transmitted from the BS. In this work, the authors first explored the relationship between the ASP and the access control at each device. Then, with a constraint on successful access probability, the primal-dual decomposition has been used to decompose the Bellman equations into multiple fixed point equations at each MTC device, which helps in making the decision at the MTC device itself [45]. It has been shown through simulation results that the proposed algorithm has outperformed all the benchmark models [45]. An LSTM-based ACB scheme has been proposed in [46] to adjust the ACB based on the predicted upcoming traffic load. It has been shown through results that the proposed mechanism achieves a 98% success rate. In addition, there is a reduction of 6% in the number of access attempts with the proposed mechanism when compared to the dynamic ACB mechanism and a drop of 68% has been observed in the access delay when compared to the reinforcement learning-based ACB scheme [46]. In [47], a novel learning with a finite memory framework has been proposed to decrease the latency of critical alarm messages in the presence of delay-tolerant and periodic messages. In the proposed framework, each MTC device adjusts the transmission parameters by learning the message states of other devices. Through simulations, it has been shown that the proposed framework lowers the latency of the critical alarm messages by up to 94% [47]. The proposed model in [47] results in higher performance with lower memory which is a green learning model. To further decrease the latency of the critical alarm messages, a novel finite-memory multistate sequential learning framework has been proposed in [48].

A novel sleeping multi-armed bandit framework has been proposed in [49] to optimally schedule the uplink grant for MTC devices. In this framework, the combination of data rate, maximum tolerable access delay, and value of data for each MTC device is considered for the reward. This reward has been used in the proposed framework to find the best MTC devices for scheduling at each time [49]. A DL approach based on LSTM has been proposed in [50] to predict event-driven traffic based on past transmission data. The proposed DL approach infers the causal relationship between the devices, which is used to predict the event-driven source traffic. It has been shown that the proposed model saves resources and improves accuracy with a margin of around 9% as compared to the existing baseline models [50]. The proposed LSTM-based model in [50] used a low amount of signaling which is a part of green AI. An LSTM-based recurrent neural network machine learning algorithm has been proposed in [51] to predict the successful preamble transmissions at a BS which will be broadcasted to control the congestion under bursty traffic conditions. In [52], the authors have proposed a decoupling learning strategy for access control optimization that works on the basis of two strategies, such as traffic prediction in the current transmission and access control configuration for future transmissions. The proposed model in [52] needs to be improved such that it controls the RACH in a cooperative manner which reduces the computational complexity to make it a green AI model. Further, an ML-based random access (RA) prediction strategy has been proposed in [53] that predicts the RA outcomes with high accuracy and drives the devices toward optimized RA configurations in order to reduce power consumption. A deep reinforcement learning model has been considered in [54] to determine the strategic parameters for controlling the number of collisions and access latency. In [55], a Q-learning-based ACB control approach has been incorporated into each mM2M device to avoid collisions in the network without any assistance from the BS. When the learning algorithm reaches convergence, the device is assigned a unique time slot for transmission to avoid network collisions. In contrast to the Q-learning approach, the collaborative Q-learning algorithm has been considered in [56] in order to learn the opponent's beliefs and environmental knowledge in order to optimize the device's waiting time.

7.3.5 Backoff mechanism

The exponential backoff is usually considered for the random access mechanism to increase the throughput and reduce the delay. However, this backoff results in varying performance under different assumptions and analytical models. Thus, we need to consider using the learning-based transmission policy to increase the throughput [57].

In [57], a deep reinforcement learning algorithm has been proposed for RA to balance the throughput and fairness. The learning probabilities of the proposed algorithm are predicted based on the action in the previous step and the binary feedback signal. The proposed algorithm is also adaptable to various rates of traffic arrival. Moreover, the average age of a packet has been proposed as a metric to estimate the fairness among the devices. It has been concluded through simulations that the proposed algorithm performs better when compared to the exponential backoff in terms of throughput and fairness [57].

7.3.6 CSI-based algorithms

The conventional methods for MTC result in poor performance due to the avoidance of channel state information. In [58], an energy-aware matching-based channel selection algorithm named E-MGSI is augmented with E-MUCB, which is a matching-learning-based algorithm to learn the unavailable CSI. Furthermore, it has been proved theoretically that the proposed E-MUCB learning-based algorithm results in better performance in comparison to the E-MGSI with global state information [58].

7.4 EXPERIMENTS

In this section, we present the numerical results corresponding to the average computing time for multiuser detection. We compare the performance of block-restrict neural network architecture which is a fast learning method proposed in [35] with non-ML based approaches (iterative sparse algorithms) such as orthogonal matching pursuit (OMP) proposed in [59], Iterative Hard Thresholding (IHT) proposed in [60], Block IHT proposed in [61] and Block OMP proposed in [62] which are the block structure extensions to IHT.

For the experimental evaluation, a total of K = 100 users are considered, and Ns pilot symbols considered for active user transmit power with binary phase shift keying (BPSK) modulation. The channel is modeled by L = 6 independent identically Rayleigh distributed taps. Noise follows Gaussian distribution with zero mean and variance is adjusted so as to obtain the desired value of signal-to-noise ratio (SNR). A total of 8×10^6 are generated for training, 10^5 samples are considered for testing and verification.

Table 7.1 Average computing time for MUD

(N_s, n)	*DNN/BRNN*	*OMP*	*IHT*	*BIHT*	*BOMP*
(30,3)	2.57×10^{-4}	0.007	0.162	0.173	0.007
(30,6)	2.57×10^{-4}	0.007	0.009	0.163	0.009
(40,3)	2.56×10^{-4}	0.009	0.183	0.159	0.006
(40,6)	2.56×10^{-4}	0.010	0.021	0.177	0.010
(50,3)	2.58×10^{-4}	0.006	0.193	0.141	0.006
(50,6)	2.58×10^{-4}	0.012	0.085	0.183	0.012

For all samples, 10 dB is considered as SNR. We activate 6 nodes randomly for training and verification and $n \leq 6$ nodes are randomly activated for testing. A learning rate of 0.01 and momentum of 0.9 are considered for training neural networks.

Table 7.1 describes the average computing time for testing one data sample for different pilot symbols $N_s = \{30, 40, 50\}$ with the number of active users $n = \{3, 6\}$. For a better comparison, we consider a computer with Lenovo Thinkpad Carbon X1 Computer with i7 processor of 2.80 GHz CPU and 32 GB RAM running on windows operating system. It is observed from Table 7.1 that the DNN/BRNN decreases the computing time by 10 times as compared to the non-ML-based approaches.

7.5 CONCLUSION

In this chapter, we have discussed the message exchange in the random access channel mechanism for massive machine-to-machine communications in cellular Internet-of-Things. Then, we have described various green machine learning models for massive machine-to-machine communication. Through experimental results, it has been shown that the ML-based approaches result in better performance when compared with non-ML-based approaches.

7.6 FUTURE DIRECTIONS

As mentioned in [15] the computations required for running a neural network model has increased by 300,000 times in six years. These computations increase the financial cost as well as the carbon emission. Further, researchers are increasing the model size for training large data so as to generate human-like results. However, training of the model will release carbon that is equal to the emitted carbon from five cars during their lifetime.

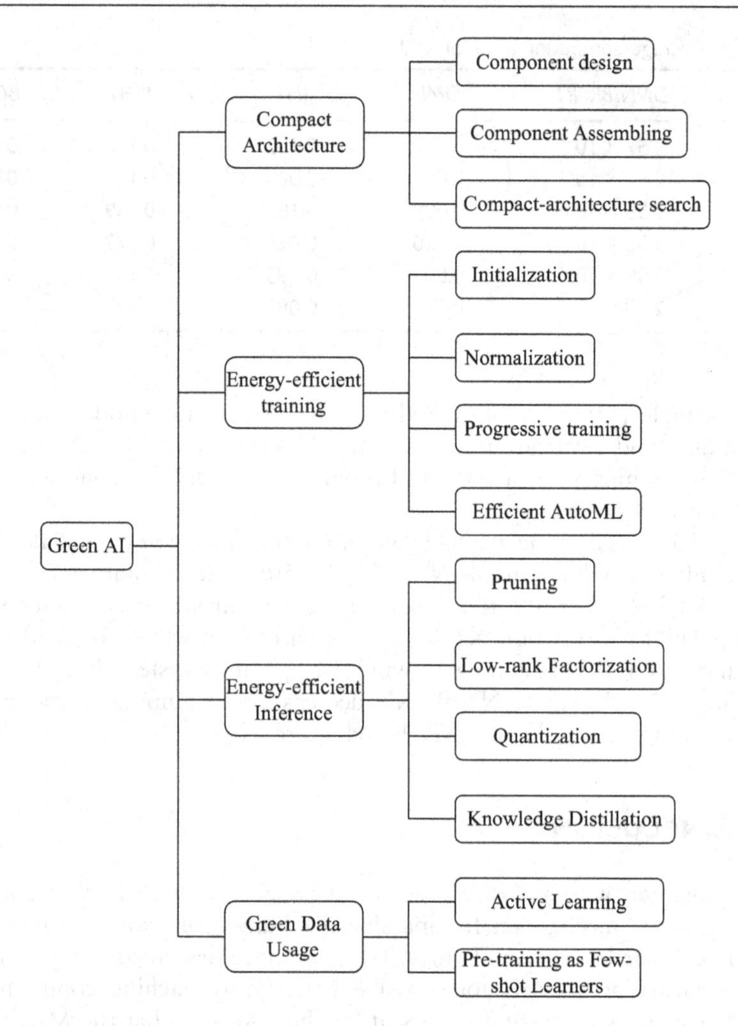

Figure 7.4 Design parameters for Green AI.

The development of efficient neural networks is a long-term goal of Green AI. Figure 7.4 lists the methods to achieve green AI by reducing the computational cost. Compact architecture, energy-efficient training, energy-efficient inference, and green data usage are the main components for obtaining green AI.

7.6.1 Compact architecture

It involves the component design and assembling and compact-architecture search which are described below:

- **Component Design:** The popular components for the design are convolution, softmax, attention, and embedding.
- **Component Assembling:** Sharing is the main idea behind efficient component assembling. Sharing of memory, static weight, dynamic weight, and deployment weight are the four categories of sharing.
- **Compact-architecture search:** In addition to the model design, searching for efficient networks for resource-constrained networks also need to be considered.

7.6.2 Energy-efficient training

It involves parameter initialization, normalization, progressive training, and efficient AutoML:

- **Initialization:** Architecture design and parameter selection are the initial steps of the training. Efficient parameter selection is one of the primary requirements for green AI.
- **Normalization:** In addition to initialization, normalization also accelerates the training. It is a special component that accelerates the convergence which must be considered for the design of green AI.
- **Progressive training:** It adds the layers constructively. In contrast to full training, progressive training does not require full gradients to all parameters which reduces the computational time for the training.
- **Efficient AutoML:** Optimization of hyper-parameters is one of the fundamental and common steps in the design of a better model. Due to the uncertainty in the tuning of hyper-parameters, there is a need for high computations for tuning them. Thus, the tuning of hyper-parameters or autoML is a field for study.

7.6.3 Energy-efficient inference

In this section, we describe the methods for reducing the inference costs such as pruning, low-rank factorization, quantization, and knowledge distillation:

- **Model pruning:** Pruning is an efficient solution to reduce redundant parameters. It has been shown in [63] that the redundant parameters are not important for the final performance. Thus, reducing the redundant parameters in the model is one of the solutions to be considered for green AI.
- **Low-rank factorization:** Tensor operation is the basic building block that contributes much to the computational cost. Thus, reducing the tensor operations is one of the methods for green AI.

- **Quantization:** Quantization is used for compressing the network by reducing the number of bits. Thus, quantization needs to be considered for achieving green AI.
- **Knowledge distillation:** Knowledge distillation exploits the knowledge in the training a student by an already trained teacher. Thus, exploiting the knowledge distillation leads to green AI.

7.6.4 Green data usage

This section mainly exploits the ways to achieve efficient results with fewer data resources:

- **Active learning:** It is an approach to obtain competitive results with few samples. By reducing the samples, the computational time is also reduced. Thus, exploiting active learning leads to green AI.
- **Pre-training as Few-shot Learners:** Considering the pre-trained models as the initialization is an effective way to reduce the computational time for green AI.

REFERENCES

[1] A. Al-Fuqaha, M. Guizani, M. Mohammadi, M. Aledhari, and M. Ayyash, "Internet of Things: A survey on enabling technologies, protocols, and applications," *IEEE Communications Surveys & Tutorials*, vol. 17, no. 4, pp. 2347–2376, Jun. 2015, doi: 10.1109/COMST.2015.2444095.

[2] H. Wang and A. O. Fapojuwo, "A survey of enabling technologies of low power and long range Machine-to-Machine communications," *IEEE Communications Surveys & Tutorials*, vol. 19, no. 4, pp. 2621–2639, Jun. 2017, doi: 10.1109/COMST.2017.2721379.

[3] "IMT Vision – Framework and overall objectives of the future development of IMT for 2020 and beyond M Series Mobile, radiodetermination, amateur and related satellite services," *International Telecommunications Union*, Geneva, Switzerland, ITU-Recommendation M.2083-0, Sep. 2015.

[4] B. A. Jayawickrama, Y. He, E. Dutkiewicz, and M. D. Mueck, "Scalable spectrum access system for massive machine type communication," *IEEE Network*, vol. 32, no. 3, pp. 154–160, May/Jun. 2018, doi: 10.1109/MNET.2017.1700149.

[5] 3GPP, "Study on scenarios and requirements for next generation access technologies (release 14)," *Tech. Rep. 38.913*, v14.2.0, pp. 1–41, Jun. 2017.

[6] Z. Dawy, W. Saad, A. Ghosh, J. G. Andrews, and E. Yaacoub, "Toward massive machine type cellular communications," *IEEE Wireless Communications*, vol. 24, no. 1, pp. 120–128, Feb. 2017, doi: 10.1109/MWC.2016.1500284WC.

[7] K. Wang, Y. Wang, Y. Sun, S. Guo, and J. Wu, "Green industrial Internet of Things architecture: An energy-efficient perspective," *IEEE Communications Magazine*, vol. 54, no. 12, pp. 48–54, Dec. 2016, doi: 10.1109/MCOM.2016.1600399CM.

[8] S. K. Sharma and X. Wang, "Distributed caching enabled peak traffic reduction in ultra-dense IoT networks," *IEEE Communications Letters*, vol. 22, no. 6, pp. 1252–1255, Jun. 2018, doi: 10.1109/LCOMM.2018.2810211.

[9] C. Bockelmann et al., "Massive machine-type communications in 5g: Physical and MAC-layer solutions," *IEEE Communications Magazine*, vol. 54, no. 9, pp. 59–65, Sep. 2016, doi: 10.1109/MCOM.2016.7565189.

[10] Y. S. Reddy, A. Dubey, A. Kumar, and T. Panigrahi, "A probabilistic approach to model SIC based RACH mechanism for machine type communications in cellular networks," *IEEE Transactions on Vehicular Technology*, vol. 70, no. 2, pp. 1878–1893, Feb. 2021, doi: 10.1109/TVT.2021.3055286.

[11] Y. S. Reddy, A. Dubey, A. Kumar, and T. Panigrahi, "A successive interference cancellation based random access channel mechanism for machine-to-machine communications in cellular Internet-of-Things," *IEEE Access*, vol. 9, pp. 8367–8380, Jan. 2021, doi: 10.1109/ACCESS.2021.3049439.

[12] A. Kulkarni and A. Shivananda, *Natural language processing recipes*, Apress, Jul. 2019.

[13] S. Bianco, M. Buzzelli, G. Ciocca, and R. Schettini, "Neural architecture search for image saliency fusion," *Information Fusion*, vol. 57, pp. 89–101, May 2020.

[14] E. Strubell, A. Ganesh, and A. McCallum, "Energy and policy considerations for deep learning in NLP," in *Proceedings of the 57th Annual Meeting of the Association for Computational Linguistics*, Florence, Italy, Jul. 2019, pp. 3645–3650.

[15] R. Schwartz, J. Dodge, N. A. Smith, and O. Etzioni, "Green AI," *Communications of the ACM*, vol. 63, no. 12, pp. 54–63, Nov. 2020.

[16] A. Imran, A. Zoha, and A. Abu-Dayya, "Challenges in 5G: How to empower SoN with big data for enabling 5G," *IEEE Network*, vol. 28, no. 6, pp. 27–33, Nov/Dec. 2014, doi: 10.1109/MNET.2014.6963801.

[17] R. Li et al., "Intelligent 5G: When cellular networks meet artificial intelligence," *IEEE Wireless Communications*, vol. 24, no. 5, pp. 175–183, Oct. 2017, doi: 10.1109/MWC.2017.1600304WC.

[18] R. C. Daniels and R. W. Heath, "An online learning framework for link adaptation in wireless networks," in *Proceedings of Information Theory and Applications Workshop*, La Jolla, CA, USA, Feb. 2009, pp. 138–140, doi: 10.1109/ITA.2009.5044935.

[19] S. Yun and C. Caramanis, "Reinforcement learning for link adaptation in MIMO-OFDM wireless systems," in *Proceedings of IEEE Global Telecommunications Conference (GLOBECOM)*, Miami, FL, USA, Dec. 2010, pp. 1–5, doi: 10.1109/GLOCOM.2010.5683371.

[20] J. Wu, I. Bisio, C. Gniady, E. Hossain, M. Valla, and H. Li, "Context-aware networking and communications: Part 1 [guest editorial]," *IEEE Communications Magazine*, vol. 52, no. 6, pp. 14–15, Jun. 2014, doi: 10.1109/MCOM.2014.6829939.

[21] S. K. Sharma and X. Wang, "Live data analytics with collaborative edge and cloud processing in wireless IoT networks," *IEEE Access*, vol. 5, pp. 4621–4635, Mar. 2017, doi: 10.1109/ACCESS.2017.2682640.

[22] L. M. Bello, P. Mitchell, and D. Grace, "Application of Q-learning for RACH access to support M2M traffic over a cellular network," in *Proceedings of European Wireless Conference*, Barcelona, Spain, May 2014, pp. 1–6.

[23] M. Hasan, E. Hossain, and D. Niyato, "Random access for machine-to-machine communication in LTE-advanced networks: Issues and approaches," *IEEE Communications Magazine*, vol. 51, no. 6, pp. 86–93, Jun. 2013.
[24] T. Park and W. Saad, "Resource allocation and coordination for critical messages using finite memory learning," in *Proceedings of IEEE Globecom Workshops (GC Wkshps)*, Washington, DC, USA, Dec. 2016, pp. 1–6, doi: 10.1109/GLOCOMW.2016.7849023.
[25] K. Portelli and C. Anagnostopoulos, "Leveraging edge computing through collaborative machine learning," in *Proceedings of International Conference on Future Internet of Things and Cloud Workshops (FiCloudW)*, Prague, Czech Republic, Aug. 2017, pp. 164–169, doi: 10.1109/FiCloudW.2017.72.
[26] 3GPP, "E-UTRA-Medium access control (MAC) protocol specification," *Tech. Rep. 36.321*, v14.3.0, pp. 1–17, Jun. 2017.
[27] Y. S. Reddy, A. Dubey, and A. Kumar, "A novel RACH mechanism for machine type communications in cellular networks," in *Proceedings of IEEE International Conference on Advanced Networks and Telecommunications Systems (ANTS)*, Bhubaneswar, India, Dec. 2017, pp. 1–6, doi: 10.1109/ANTS.2017.8384154.
[28] N. Jiang, Y. Deng, and A. Nallanathan, "Traffic prediction and random access control optimization: Learning and non-learning-based approaches," *IEEE Communications Magazine*, vol. 59, no. 3, pp. 16–22, Mar. 2021.
[29] El-Hameed, A.S.A. and Elsayed, K.M.F., "A Q-learning approach for machine-type communication random access in LTE-Advanced," *Telecommun Systems*, vol. 71, pp. 397–413, Jul. 2019. https://doi.org/10.1007/s11235-018-0509-2
[30] A. R. Rosyadi, T. A. B. Wirayuda, and S. Al-Faraby, "Intelligent traffic light control using collaborative Q-Learning algorithms," in *Proceedings of International Conference on Information and Communication Technology (ICoICT)*, Bandung, Indonesia, May 2016, pp. 1–6.
[31] M. Hasan, E. Hossain, and D. Niyato, "Random access for machine-to-machine communication in LTE-advanced networks: issues and approaches," *IEEE Communications Magazine*, vol. 51, no. 6, pp. 86–93, Jun. 2013, doi: 10.1109/MCOM.2013.6525600.
[32] A. H. Mohammed, A. S. Khwaja, A. Anpalagan, and I. Woungang, "Base station selection in M2M communication using Q-learning algorithm in LTE-A networks," in *Proceedings of IEEE International Conference on Advanced Information Networking and Applications*, Gwangju, Korea (South), Mar. 2015, pp. 17–22, doi: 10.1109/AINA.2015.160.
[33] Y. -J. Liu, S. -M. Cheng, and Y. -L. Hsueh, "eNB selection for machine type communications using reinforcement learning based Markov decision process," *IEEE Transactions on Vehicular Technology*, vol. 66, no. 12, pp. 11330–11338, Dec. 2017, doi: 10.1109/TVT.2017.2730230.
[34] N. Kim, D. Kim, B. Shim, and K. B. Lee, "Deep learning-based spreading sequence design and active user detection for massive Machine-Type Communications," *IEEE Wireless Communications Letters*, vol. 10, no. 8, pp. 1618–1622, Aug. 2021, doi: 10.1109/LWC.2021.3071453.
[35] Y. Bai, B. Ai, and W. Chen, "Deep learning based fast multiuser detection for massive machine-type communication," in *Proceedings of IEEE Vehicular Technology Conference (VTC2019-Fall)*, Honolulu, HI, USA, Sep. 2019, pp. 1–5, doi: 10.1109/VTCFall.2019.8891371.

[36] B. Liu, Z. Wei, W. Yuan, J. Yuan, and M. Pajovic, "Channel estimation and user identification with deep learning for massive machine-type communications," *IEEE Transactions on Vehicular Technology*, vol. 70, no. 10, pp. 10709–10722, Oct. 2021, doi: 10.1109/TVT.2021.3111081.
[37] Y. Ahn, W. Kim, and B. Shim, "Active user detection and channel estimation for massive machine-type communication: Deep learning approach," *IEEE Internet of Things Journal*, early access, Dec. 2021, doi: 10.1109/JIOT.2021.3132329.
[38] L. M. Bello, P. Mitchell, and D. Grace, "Application of Q-learning for RACH access to support M2M traffic over a cellular network," in *Proceedings of European Wireless Conference*, Barcelona, Spain, May 2014, pp. 1–6.
[39] Y. Ruan, W. Wang, Z. Zhang, and V. K. N. Lau, "Delay-aware massive random access for machine-type communications via hierarchical stochastic learning," in *Proceedings of IEEE International Conference on Communications (ICC)*, Paris, France, May 2017, pp. 1–6, doi: 10.1109/ICC.2017.7996795.
[40] J. Moon and Y. Lim, "Access control of MTC devices using reinforcement learning approach," in *Proceedings of International Conference on Information Networking (ICOIN)*, Da Nang, Jan. 2017, pp. 641–643, doi: 10.1109/ICOIN.2017.7899576.
[41] W. Abera, T. Olwal, Y. Marye, and A. Abebe, "Learning based access class barring for massive machine type communication random access congestion control in LTE-A networks," in *Proceedings of International Conference on Electrical, Computer and Energy Technologies (ICECET)*, Cape Town, South Africa, Dec. 2021, pp. 1–7, doi: 10.1109/ICECET52533.2021.9698652.
[42] B. Sliwa, T. Liebig, R. Falkenberg, J. Pillmann, and C. Wietfeld, "Efficient machine-type communication using multi-metric context-awareness for cars used as mobile sensors in upcoming 5G networks," in *Proceedings of IEEE Vehicular Technology Conference (VTC Spring)*, Porto, Portugal, Jun. 2018, pp. 1–6, doi: 10.1109/VTCSpring.2018.8417753.
[43] Z. Chen and D. B. Smith, "Heterogeneous machine-type communications in cellular networks: Random access optimization by deep reinforcement learning," in *Proceedings of IEEE International Conference on Communications (ICC)*, Kansas City, MO, USA, May 2018, pp. 1–6, doi: 10.1109/ICC.2018.8422775.
[44] M. V. da Silva, R. D. Souza, H. Alves, and T. Abrão, "A NOMA-based Q-learning random access method for machine type communications," *IEEE Wireless Communications Letters*, vol. 9, no. 10, pp. 1720–1724, Oct. 2020, doi: 10.1109/LWC.2020.3002691.
[45] Y. Ruan, W. Wang, Z. Zhang, and V. K. N. Lau, "Delay-aware massive random access for machine-type communications via hierarchical stochastic learning," in *Proceedings of IEEE International Conference on Communications (ICC)*, Paris, France, May 2017, pp. 1–6, doi: 10.1109/ICC.2017.7996795.
[46] Chu-Heng Lee, Shang-Juh Kao, Fu-Min Chang, "LSTM-based ACB scheme for machine type communications in LTE-A networks," *Computer Communications*, vol. 152, Feb. 2020, pp. 296–304. doi: 10.1016/j.comcom.2020.01.047.
[47] T. Park and W. Saad, "Learning with finite memory for machine type communication," in *Proceedings of Annual Conference on Information Science and Systems (CISS)*, Princeton, NJ, USA, Mar. 2016, pp. 608–613, doi: 10.1109/CISS.2016.7460572.

[48] T. Park and W. Saad, "Distributed learning for low latency machine type communication in a massive Internet of Things," *IEEE Internet of Things Journal*, vol. 6, no. 3, pp. 5562–5576, Jun. 2019, doi: 10.1109/JIOT.2019.2903832.
[49] S. Ali, A. Ferdowsi, W. Saad, N. Rajatheva, and J. Haapola, "Sleeping multi-armed bandit learning for fast uplink grant allocation in machine type communications," *IEEE Transactions on Communications*, vol. 68, no. 8, pp. 5072–5086, Aug. 2020, doi: 10.1109/TCOMM.2020.2989338.
[50] T. Senevirathna, B. Thennakoon, T. Sankalpa, C. Seneviratne, S. Ali, and N. Rajatheva, "Event-driven source traffic prediction in machine-type communications using LSTM networks," in *Proceedings of IEEE Global Communications Conference*, Taipei, Taiwan, Dec. 2020, pp. 1–6, doi: 10.1109/GLOBECOM42002.2020.9322417.
[51] A. Søraa, T. N. Weerasinghe, I. A. M. Balapuwaduge, and F. Y. Li, "Preamble transmission prediction for mMTC bursty traffic: A machine learning based approach," in *Proceedings of IEEE Global Communications Conference*, Taipei, Taiwan, Dec. 2020, pp. 1–6, doi: 10.1109/GLOBECOM42002.2020.9322633.
[52] N. Jiang, Y. Deng, A. Nallanathan, and J. Yuan, "A decoupled learning strategy for massive access optimization in cellular IoT networks," *IEEE Journal on Selected Areas in Communications*, vol. 39, no. 3, pp. 668–685, Mar. 2021, doi: 10.1109/JSAC.2020.3018806.
[53] G. Caso, K. Kousias, Ö. Alay, A. Brunstrom, and M. Neri, "NB-IoT random access: Data-driven analysis and ML-based enhancements," *IEEE Internet of Things Journal*, vol. 8, no. 14, pp. 11384–11399, Jul. 2021, doi: 10.1109/JIOT.2021.3051755.
[54] Jihun Moon and Yujin Lim, "A reinforcement learning approach to access management in wireless cellular networks," *Wireless Communications and Mobile Computing*, vol. 2017, pp. 1–8, May 2017.
[55] S. K. Sharma and X. Wang, "Toward massive machine type communications in ultra-dense cellular IoT networks: Current issues and machine learning-assisted solutions," *IEEE Communications Surveys & Tutorials*, vol. 22, no. 1, pp. 426–471, first quarter 2020.
[56] C. Lamini, Y. Fathi, and S. Benhlima, "Collaborative Q-learning path planning for autonomous robots based on holonic multi-agent system," in *Proceedings of International Conference on Intelligent Systems: Theories and Applications (SITA)*, Rabat, Morocco, Oct. 2015, pp. 1–6.
[57] M. A. Jadoon, A. Pastore, M. Navarro, and F. Perez-Cruz, "Deep reinforcement learning for random access in Machine-type communication," in *Proceedings of IEEE Wireless Communications and Networking Conference (WCNC)*, Austin, TX, USA, Apr. 2022, pp. 2553–2558, doi: 10.1109/WCNC51071.2022.9771953.
[58] H. Liao and Z. Zhou, "Learning-based energy-aware channel selection for machine type communications," in *Proceedings of IEEE Globecom Workshops (GC Wkshps)*, Waikoloa, HI, USA, Dec. 2019, pp. 1–5, doi: 10.1109/GCWkshps45667.2019.9024505.
[59] D. L. Donoho, Y. Tsaig, I. Drori, and J. Starck, "Sparse solution of underdetermined systems of linear equations by stagewise orthogonal matching pursuit," *IEEE Transactions on Information Theory*, vol. 58, no. 2, pp. 1094–1121, Feb. 2012.

[60] R. E. Carrillo, L. F. Polanía, and K. E. Barner, "Iterative hard thresholding for compressed sensing with partially known support," in *Proceedings of IEEE International Conference on Acoustics, Speech and Signal Processing (ICASSP)*, Prague, Czech Republic, May 2011, pp. 4028–4031.
[61] R. Garg and R. Khandekar, "Block-sparse solutions using kernel block rip and its application to group lasso," in *Proceedings of the Fourteenth International Conference on Artificial Intelligence and Statistics*, Jun. 2011, pp. 296–304.
[62] Y. Fu, H. Li, Q. Zhang, and J. Zou, "Block-sparse recovery via redundant block omp," *Signal Processing*, vol. 97, pp. 162–171, Apr. 2014.
[63] S. Hanson and P. Lorien, "Comparing biases for minimal network construction with back-propagation," in *Proceedings of Advances in neural information processing systems 1, NIPS Conference*, Denver, Colorado, USA, 1988, pp. 177–185.

Index

Pages in *italics* refer to figures.

Printed in the United States
by Baker & Taylor Publisher Services

Printed in the United States
by Baker & Taylor Publisher Services